The Animals of Grandfather Mountain

An Animal Caretaker's Tales

"My Best Boy, Mumbles"

REVISED EDITION

by

L. L. Mitchell

1st Edition published by Parkway Publishers Inc., Boone, NC -- 2001

Revised Edition published by L. L. Mitchell, Dubois, WY -- 2011

Library of Congress Control Number: 2011912334

ISBN-13: 978-1463715861

Hugh Morton, former owner and president of Grandfather Mountain, took many photos on the following pages of this book. I'm very honored that he gave me permission to use them. He was an avid conservationist, a professional wildlife photographer and a great man who carved out an important and unique part of North Carolina history. He loved Grandfather Mountain with all his heart and wanted to share that love with others.

He will never be forgotten by what he achieved in making Grandfather Mountain such a special place to educate the public about nature.

_____

Cover Photo by L. L. Mitchell

Other photos in this book were taken by L. L. Mitchell unless otherwise noted.

Habitat Map Illustration by Jim Fleri

To order additional copies of this book, visit the author's website at:

www.llmitchell.com

or

www.theanimalsofgrandfathermountain.com

This book is dedicated to the memory of my granddaddy,

John Rayford Freeman

Who was Mildred the Bear?   Photo by Hugh Morton

In 1968, a local wildlife club proposed a program for the repopulating of black bears in the North Carolina Mountains. Grandfather was asked to purchase two bears for release into the backcountry of the mountain, a protected wildlife area. A male and female bear were purchased from the Atlanta Zoo. The male ran away immediately after release, but the female bear was in no hurry to leave her human companions. When the female was released, she was photographed and filmed for the Arthur Smith television show so a song could be edited to the bear's departure. The problem was she didn't want to depart!

She stayed with the film crew all day, and when they left, she felt abandoned and went in search of human company in Linville. The NC wildlife officials contacted Grandfather Mountain and insisted that this nuisance bear be recaptured and put back in her cage! It was later learned that the female bear, who comedian Ralph Smith named "Mildred" the day of the television show filming, had been hand-raised by the secretarial staff at the Atlanta Zoo. She was entirely too trusting of humans to be released into the wild. A beautiful two-acre habitat was built for Mildred and her family in 1973. Mildred gave birth to ten cubs over the years and adopted three that were not her own, very unusual behavior for black bears. Mildred died in her sleep at the age of 26 on January 1, 1993. She is greatly missed by all the staff and visitors who had grown to know and love her throughout the years.

On the following pages, you will read about and see photos of animal caretakers in enclosures with black bears. These keepers were trained to be careful and watch the bears for signs of agitation. In recent years, holding cages have been installed so keepers and bears are no longer in direct contact when keepers enter the enclosures.

Grandfather's bears may seem tame to visitors, but they still maintain their wild instincts. Never approach or feed a wild bear or any other wild animal no matter how tame it might seem.

Please remember:
Wild animals do not make good pets

FOREWORD
by Hugh Morton
(1921 - 2006)

Hugh Morton and his beloved friend, Mildred the Bear
(photo used with permission -- Grandfather Mountain, Inc.)

Laurie Mitchell came to Grandfather Mountain as a graduate of Auburn University with a degree in Wildlife Biology, and previous experience working with animals at the Montgomery Zoo in Alabama. She left Grandfather Mountain with a wide circle of new friends, and genuine love for bears, deer, cougars, eagles and otters. Her story could be the basis for a television or Hollywood production, but if that happens it will have to be later; for now she has authored an engaging book that is our pleasure to read.

The environmental habitats for native animals at Grandfather that Laurie gave such tender love and care are a real challenge. They are larger than the habitats in some of the nation's best known zoos, and this happily gives Grandfather's animals plenty of space for their homes. On the down side, the particularly private animals like Cougars sometimes are out of view of the Grandfather Mountain visitors. Some may say that rocket science is more involved and difficult. We maintain that the skill and psychology shown by Laurie and her staff in caring for the animals, yet always remembering the pleasure of the visitors by working to assure that they see what they came to see, takes talent and dedication that compares with any profession.

The habitats at Grandfather Mountain will always be named for Mildred the Bear, the nicest bear that has ever been. When Grandfather agreed to obtain two bears to be released in the wild to help rebuild the bear population in the mountains, by mistake the Atlanta Zoo sold Grandfather a young bear that the office staff at the zoo had raised on a bottle.

Consequently, when that young bear was released, she did not revert to the wild. She did not know she was a bear. She had none of the hostility sometimes associated with bears; she just wanted to hang out with people. She was given the name Mildred.

Faced with the problem of providing a home for a friendly bear that would not turn wild, Grandfather Mountain obtained the expert advice of Bill Hoff, then Director of the North Carolina Zoo at Asheboro, and J. Hyatt Hammond, the architect who had done much of the design work for the state zoo. Taking full advantage of the natural terrain, Hoff and Hammond designed the original habitat that nestles between giant boulders. We have been told time and again that what they designed is the best display for Black Bears in the world. Hammond's architectural firm later designed the building that houses the Grandfather Mountain Nature Museum.

Habitats for White-tailed Deer, Cougars (also known as Panthers and Mountain Lions), Bald Eagles, Golden Eagles, and River Otters followed. A habitat for bear cubs was also needed, because the only animal friend a cub has in the world is its own mother. Laurie was given a free hand to select her own staff for the habitats. The area was always neat. The animals were being fed the best known diets for their respective species. If any animal appeared under the weather, it received immediate attention from veterinarians who were always on call. Laurie's book, as well as the immaculately kept habitats, is proof that a remarkable lady loved her job.

CONTENTS

Laurie with Gerry the Bear   Photo by Hugh Morton

Preface

From the time I can remember, I wanted to work with animals. My granddaddy instilled in me a love for animals and nature from the time I was a small girl. If it had not been for him, I would never have gotten the chance to walk among bears.

I first saw Grandfather Mountain in early March 1991 when I was visiting the High Country for the first time. He was covered with snow and was a very magnificent fellow indeed! It was also the first time I glimpsed the animals in the natural habitats. The bears were asleep, of course, but the others were stirring. The deer were hunkered down chewing cud, and the eagles sat on their perches, fussing and chattering. One of the cougars must have been cold because he was pacing all around his snow-filled exhibit. Looking at that cat that day, I never thought that I would actually know him on a personal basis.

Wouldn't it be wonderful to work in a majestic place such as this? I thought. Never did I imagine, as I left the park, that Grandfather Mountain had a plan for me to do just that.

Six years later, the Mountain came calling. I was a zookeeper at the City of Montgomery Zoo in Alabama at the time. But I didn't want to be in Alabama. I wanted to be in the mountains ... the North Carolina Mountains. And in September 1996, I was offered the job as Animal Habitat Manager overlooking those animals I had visited so long ago ...

During the four years I spent at Grandfather Mountain taking care of the animals, I also made many human friends as well. Kirsten, the assistant habitat manager, and I began working in the habitats around the same time. The habitat staff had gone through a change that first September, and both of us were new. And as the seasons changed, we learned the different tasks that awaited us.

When we began our employment in the fall, the leaves were doing their duty. And it became our duty to rake leaves and clean ponds of those leaves. That fall we were also introduced to the many school groups that visited Grandfather Mountain and learned rather quickly how to give tours.

As winter approached, we learned to use the snow blower, and to salt and shovel the walks when the snow began to fall. In the spring, with more school groups visiting, we hired new seasonal employees to help us. Throughout the summer, the staff cleaned ponds, painted rails, and animal dens, and mowed and trimmed grass in addition to feeding and taking care of the animals. The habitat staff prided itself in keeping the grounds first rate.

But the main reason we did these sometimes very physical and exhausting chores was because we loved our animals. We wanted them to have the best habitats and to be as healthy as possible. We wanted to share with people how special nature and its animals are. Talking with visitors in the habitats about the animals was a very important endeavor.

The main stars in the habitats surely aren't the animal keepers. But I would like to thank all the habitat employees who worked for me during my years at Grandfather Mountain. It is with pride that I thank you for a job well done.

I must also thank each and every habitat employee who is part of the stories--Kirsten Bartlow, Tanya Alley, Pam Scarborough, Rene' Loomis, Beth Garrison, Sherri Wesley, Dana Drenzek, and Mike Wagner. Thanks also goes to the animals' veterinarian Howard Johnson, Jr., a wonderful vet. Without these people (and the animals) no book could have been written.

Other mountain employees who had a part in the stories are John Church, Richard Brown, Martha Oberhelman, Steve Miller, Harry and Kathleen Lowman, Hugh Morton, and Harris Prevost.

A special thanks goes to Hugh Morton, owner and president of Grandfather Mountain, who took time out of his very busy schedule to write the foreword for *The Animals of Grandfather Mountain* and generously supplied many of the photographs to complement the following pages. Without Mr. Morton's vision, Grandfather Mountain would not be what it is today, a top ecotourist destination where visitors can see one of the most beautiful places on Earth. Another big thank you goes to Harris Prevost, vice-president, who saw the potential in this book and encouraged me to put it together.

I greatly enjoyed working at Grandfather Mountain with its special people and animals, and I will never forget my experiences in the habitats. It is a part of my life that I will treasure forever.

-- Laurie L. Mitchell

(L. L. Mitchell)

Spring 2001

Laurie with Grandfather's Bears - Gerry, Maxi & Mumbles   photo by Paulette Freeman

Some of the animals I worked with at Grandfather Mountain have crossed over the Rainbow Bridge since this book's first edition was published, but some are still there--with new animal friends--entertaining and educating the visitors.

However, all the animals on the following pages still live on in my heart.

Chapter 1
THE CAST OF CHARACTERS

MEET THE BEARS

Maxi, born February 6, 1970, is one of famous Mildred's original cubs. As far as we know, she is the oldest black bear ever to have lived in North Carolina. Maxi's twin sister, Mini, passed away in January 1999. Many people still can remember the song Arthur Smith sang called "Mildred, Mini and Maxi". This popular tune was about Mildred's arrival to Grandfather Mountain in 1968 and about her first cubs and their antics. These three bears started the bear legacy on Grandfather over 40 years ago.

Gerry, born in 1989, was orphaned when young. Even though researchers placed her with a wild mother, she had become too acclimated to people while they were trying to find her a permanent home. She grew up wild with her adoptive mother as part of a research project led by bear biologist, Dr. Lynn Rogers, who has been studying black bears in Minnesota for over thirty years. When the research ended, Dr. Rogers realized that Gerry could not stay in the wild due to her trusting nature with people. She came to Grandfather Mountain in 1992 at 3 1/2 years old.

Photos by L. L. Mitchell

Jane was born in 1983. The North Carolina Wildlife Resources Commission gave her to Grandfather Mountain. She was confiscated from a roadside zoo where she had been abused. She was a mighty lucky bear to be able to take her place with the others on Grandfather Mountain. She is pictured here with her cubs.

photo by L. L. Mitchell

Mumbles was born in January 1980. He grew to weigh six hundred pounds, the largest black bear that I have ever seen. Mumbles was also my favorite and such a good-natured boy.

photos by L. L. Mitchell

Carolina & Dakota, born in January 1996, were purchased from Bear Country U.S.A., a drive through Animal Park in South Dakota, and the habitat staff had the pleasure of hand feeding them from cubs. They were named for North Carolina and South Dakota, their home states, respectively.

Yonahlossee & Kodiak, born in February 1999, also came from Bear Country U.S.A. Yonahlossee, a female, was named after the Yonahlossee Road, now part of U.S. 221 running from Linville to Blowing Rock. This Cherokee word means "The Trail of the Black Bear". Kodiak, a male, is a rare cinnamon colored black bear. Less than one percent of black bears in the eastern United States are cinnamon in color. In the west, brown colored black bears are more numerous. Many people don't know that black bears come in different colors, so Kodiak is educating the public. We named him after the big brown bears of Kodiak Island, Alaska. It's a big name to fill!

Punkin & Elizabeth were born in January 1983 and 1984 respectively. Mildred adopted them. Black bears are normally hostile to cubs that are not their own, but Mildred adopted Punkin when she already had two tiny cubs, and she adopted Queen Elizabeth in a year she had no cubs at all.

MEET THE EAGLES ...

Wilma and Sam are the bald eagles; the golden eagles are **Goldie and Morely** in the stories. All four of Grandfather's eagles were injured by gunshot in the western United States and had full or partial wing amputations to save their lives. Rehabilitated eagles help to educate the public by serving as ambassadors for their species.

photos by L. L. Mitchell

MEET THE RIVER OTTERS ..

Manteo and Oconee came to Grandfather Mountain from the Bayou Otter Farm in Louisiana in the summer of 1996 with the opening of the state-of-the-art otter habitat. They were named after the town of Manteo along the outer banks of North Carolina and the Oconee River in South Carolina. They were a great hit with the visitors from the very start.

Nola and Chucky joined the other two otters in January 1997. Their names were selected from an otter-naming contest. The clever contestant named them after the Nolichucky River--very appropriate names for river otters!

photos by L. L. Mitchell

photo by Hugh Morton

MEET THE COUGARS ...

Mina was born to the first resident cougars, Terra and Rajah, in 1982. Her mother, Terra, was an endangered eastern cougar. Mina left Grandfather Mountain when small and grew up on a farm in Florida. When she was five, Mina was brought back to the habitat to take her place among the other cougars.

Sheaba and Squeak were born in 1990 and 1986. The wildlife officials confiscated them from people who had them illegally as pets.

photos by L. L. Mitchell

It is sad to say that many states do not have laws against owning big cats as pets (such as cougars, lions and tigers) and people buy these animals as cubs on the spur of the moment, thinking that owning one would be neat. But cubs aren't so cuddly when they begin destroying furniture and yards. As they grow into adulthood, they eat people out of house and home ... and they can become very unpredictable. One moment, they can be friendly, and the next, they can turn on their owner. People have been maimed or killed by their "pet" cats. No wild animal makes a good pet. You can take the animal out of the wild, but you can't take the wild out of the animal.

MEET THE WHITE-TAILED DEER ...

Heidi, Critter and Star, all born in 1986 or 1987, were found in the woods when they were fawns and, because of unfortunate circumstances, were taken away from their natural homes and raised by humans. They became imprinted, which means they lost all fear of people, and could never be released back into the wild. When a fawn is born, its long spindly legs are not strong enough to keep up with its mother. The fawn knows to lie completely still, and its spots help camouflage it so predators will not find it. The mother deer hides her baby in a clump of grass or under a bush and leaves for hours at a time as she forages for food. She always returns to feed and check on the fawn. Many people who come across a hiding fawn think the mother has abandoned it or has been killed. Chances are, however, that the fawn's mother is around and is taking very good care of her baby.

Flora, born in 1998, is Star's daughter.

Fauna & Merriweather, born in 1999, are Heidi's twin daughters. They were named after the fairies in the story about Sleeping Beauty.

The Count, born in 1998, came from Transylvania County, North Carolina, and everyone knows who comes from Transylvania ... Count Dracula! That's how he got his unusual name.

photos by L. L. Mitchell

MEET THE ANIMAL CARETAKERS

Laurie Mitchell (L. L. Mitchell) came to Grandfather Mountain in October 1996 as the animal habitat manager. She earned a degree in wildlife biology in 1989 from Auburn University and was a zookeeper at the City of Montgomery Zoo in Montgomery, Alabama before venturing to the North Carolina mountains. She moved to Wyoming in September 2000 to pursue a career in writing.

Laurie & Mumbles

photo by Hugh Morton

Kirsten with Carolina & Dakota as cubs

Kirsten Bartlow came to Grandfather Mountain in October 1996 as the assistant habitat manager. She earned her degree in middle secondary education from Kansas State University. Her love for education and wildlife steered her toward a career as an environmental educator for the National Park Service, working on the Blue Ridge Parkway before coming to Grandfather. She left Grandfather in June 1998 and obtained her master's degree in Natural Resources and Environmental Management from Ball State University in May 2000. She now works as "Watchable Wildlife Coordinator" for the Arkansas Game and Fish Commission.

Tanya Alley began her career at Grandfather Mountain in April 1997. She had previously obtained a degree in ecology and environmental biology at Appalachian State University (ASU) and wanted to pursue a career working with animals. She was promoted to assistant manager in June 1998 and became the animal habitat manager in September 2000.

Tanya and Heidi

Pam Scarborough has a B.S. in biology from ASU. She worked on the mountain from 1986-1990 as the mountain's naturalist, presenting raptor programs and nature walks and caring for the habitat animals. While absent from Grandfather, she taught high school biology for five years. From April 1998 to August 2000, she worked at Grandfather Mountain as an animal caretaker.

Pam with Yonahlossee

Rene' Loomis received a psychology degree from ASU, but after graduation, pursued a career caring for animals. She worked at Grandfather Mountain from April 1998 to September 1999. She left Grandfather Mountain to become an animal caretaker at Maymont Park in Richmond, Virginia.

Rene' as Milton the Bear

Beth Garrison arrived at Grandfather Mountain in June 1998. Working on a biology degree, she already had experience as a keeper at the Natural Science Center of Greensboro, NC. She became a keeper at the North Carolina Zoo in August 2000 upon completion of her biology degree from ASU.

Sherri & Maze the Corn Snake

Sherri Wesley joined the Grandfather Mountain Staff in August 1999. While working in the habitats, she pursued her degree in biology from ASU.

Dana Drenzek, a biology student at ASU, became an animal caretaker for Grandfather Mountain in June 2000.

Dana & Dakota the Bear

Mike Wagner began his internship at Grandfather Mountain in April 2000 and graduated from ASU in May 2000 with a degree in Recreational Management.

The Veterinarian

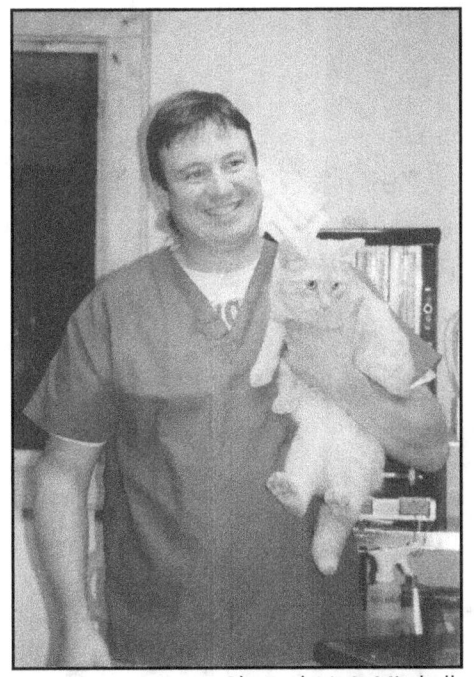

Dr. Howard Johnson, Jr. graduated from the University of Georgia with a Doctorate of Veterinary Medicine in 1988. He has been Grandfather Mountain's veterinarian, as well as a small animal vet in Boone, NC since 1989. He opened a private clinic in Boone in May 2000 while continuing to provide for the medical needs of the animals of Grandfather Mountain.

Photos by L. L. Mitchell

Other Grandfather Mountain employees who make an appearance in the stories are: **Hugh Morton**, president; **Harris Prevost**, Vice-President (even though he wasn't there!); **John Church**, maintenance manager; **Richard Brown**, assistant maintenance manager; **Harry & Kathleen Lowman**, carpenter and assistant; **Steve Miller**, trails manager; and **Martha Oberhelman**, bear hut employee.

MUMBLES & ME   PHOTO BY PAULETTE FREEMAN

CHAPTER 2

A DAY IN THE LIFE OF AN ANIMAL CARETAKER

The bald eagles see Tanya and me entering the habitat. Wilma stretches her neck like an ostrich, and her white head feathers stand on end as she peers around the rhododendron at our approach. She starts calling and Sam, her mate, joins in. They watch us warily as we pass, then settle down as we disappear down the path that leads to the bear habitats.

Mumbles the bear is lying against the front wall of the bear habitat, the only one we see in the two-acre exhibit. The others are dozing out of view. We call to Mumbles and he flicks an ear. But it isn't time for him to wake up just yet. All the bears are late risers, except for the cubs, that is. They're already running around their exhibit, climbing trees and playing chase.

Sam and Wilma   photos by L. L. Mitchell

Walking back up to the otter habitat, we pass the bald eagles again. Once more, Wilma and Sam warn everyone of our approach. The golden eagles, Goldie and Morely, just stare at us from their perch as we enter the otter house.

Manteo, Oconee, Nola and Chucky, curled up in a large four-otter ball, barely open their eyes as we peer in at them. It's probably going to be one of those water hose days to get them out onto display. But we give them the benefit of the doubt, and open the door to the exhibit to see if they will venture out on their own. Then we head to the cougar house.

The cougars greet us with little meows and purrs as we enter the cat house. Cougars are the largest cats that can purr. We walk around in their habitat before we let them out to make sure that no trees have fallen on the fence and to pick up a pair of sunglasses that somebody dropped into the habitat the day before. It is unbelievable how many objects visitors accidentally drop into the habitats.

Tanya lets the cougars out and begins cleaning dens while I feed the deer. Heidi is being a pest, as usual. I can't get around her because she has her head in the bucket, trying to eat the sweet feed and deer chow. Finally, I pull the bucket away and she follows me to the feed trough. The others hang back until I leave. I glance at The Count to check on the growth progress of his antlers. They are getting big and look like they are about to fork.

Tanya Feeds the Deer

While Tanya is still cleaning the cougar house, I walk back up to the otters to see if they're stirring. Already, visitors are looking around the display for them. I notice they are still in that otter ball "formation" inside their house.

"You'd better get out!" I say. "I'll get the hose!"

Manteo opens his eyes and peers up at me. But it's too much of an effort for him to keep them open. Slowly, his eyes become slits again.

Reluctantly, I hook the hose up and squirt water into their den. Immediately, they begin grunting and stirring. Again, I shoot a light stream of water at them. This time, I get a little more action and they slowly stretch and yawn and start up the steps, still grunting. River otters may like water, but they hate being squirted with a hose. It gets them out every time.

I clean out the otter house, then head back to the office. Tanya is already there gathering the bear food. People are amazed to find out we feed them apples, carrots, lettuce, and sweet potatoes, as well as dog food.

After we get all the food together, we set the buckets on the porch and walk inside the office. Five minutes later, I walk back out and notice that a little red squirrel has stolen one of the apples out of the bucket and is running down the road with it.

"Hey! Come back!" I yell.

The little squirrel drops the apple and hops into a tree, fussing. I pick the apple up

Dana and Jessi Feed the Bears. Kodiak is impatient!

and place it back into the bucket, shaking my head. That little "boomer" tries to steal the bears' apples all the time!

Suddenly, two chipmunks burst from the underbrush near the tool shed, one in hot pursuit of the other. They race across the road and disappear amid the leaves and twigs on the other side. I can hear the frantic chase continue around the snow blower shed and into the ditch.

Gerry the bear is waiting patiently at the gate for her apple. I slip the apple into her mouth and enter the bear habitat. We see Carolina and Dakota ambling our way and quickly lead Gerry to her feeding spot. We have to work fast so Gerry won't see the other bears. She has not been getting along with them lately.

Our efforts prove futile. Even as we lay her food out, Gerry spots Carolina and chases her out of view. Both bears grumble and blow as they run. Dakota, not the object of the chase at the moment, continues to head our way. We direct her to the other side of the habitat to be out of Gerry's wrath. As we pass the bear pond, we notice that it needs to be cleaned -- another project to add to the list.

Just then, the museum calls over the radio. "Habitats? There's a school group here that needs a tour!"

Quickly, we finish feeding the bears, and I head up to the museum to find the school group. On the way, I notice those little chipmunks again, still chasing each other through the brush. They are panting noticeably and are exhausted. In the middle of the gravel road, they stop and face each other. One punches the other, and it falls down. The accosted chipmunk gets to its feet and knocks the other down. Then they run off in opposite directions.

Still laughing at those chipmunks, I collect the school group at the museum, and we enter the habitats. I'm lining the children up in front of the deer habitat and telling the children deer facts, when another little chipmunk runs onto the path and attracts their attention. The children are amazed at the quick fragile looking little rodent. Sometimes, it seems as if visitors are more enthralled with chipmunks than cougars!

The otters, unbelievably, are cooperating beautifully in the underwater viewing area. Usually, they are sleeping in the sun, and I have a hard time luring them into the water. Today, they're swimming and playing and the children laugh at these animal comics.

The otters frolicking in their pool

"What do otters eat?" I ask, trying to stimulate the children's little minds.

"Fish!"

"That's right! And they also eat crayfish, fresh water mussels and clams and insects ... and even baby birds if they fall out of their nest," I add.

One pint-sized child raises his hand. "They might eat a baby beaver."

I raise my eyebrows and nod my head. "Yes. I guess they *would* eat a baby beaver if they came across one."

When we reach the upper otter viewing area, the otters are still in a playful frenzy. Suddenly, I hear a child exclaim, "Hey! That otter has a chipmunk!"

Sure enough, Oconee has a chipmunk in her mouth and is running crazily around while Nola chases her in a futile attempt to steal it away. Like a bullet, Oconee dives into the water, and when she comes out, the poor little drenched chipmunk, obviously dead, is still dangling precariously from her mouth.

"Oh, no!" the children cry.

Oconee stops, right in front of the horror-stricken children, and begins to eat it.

The only thing I can think to say is, "And what is something else otters eat?"

"Chipmunks!" they all yell in unison.

After several more tours, Tanya and I can finally think about cleaning the bear pond. We drain the pond, then start hosing it out. Visitors watch in shock and disbelief as Mumbles, our large male bear, walks up to the pond and stares at us.

"Aren't you scared of those bears?" a visitor asks.

"No," I answer. "These bears were either hand raised or were born here. They're used to people going in with them. If we respect them, they'll respect us."

Mumbles usually isn't the problem; Carolina and Dakota are. Sure enough, hearing the water hose, they shuffle out of the rocks and enter the pond looking for salamanders and overlooked peanuts. Getting the two bear nuisances out proves no easy task. Unlike the otters, they like

Bears Love the Hose!

being squirted with the water hose -- especially when it's hot. But somehow we get the job done.

Then there's weed trimming, mowing grass, more pond cleaning and painting. The habitat staff does most of its maintenance work, and the chores seem endless. Depending on what we're doing at the moment, visitors are either wishing they were habitat employees or are happy that they are not.

Pam offers to feed the eagles in the late afternoon. Wilma is sitting on an egg, and Pam knows to be careful. The female bald eagle has become aggressive since she laid that egg, and she won't let anyone get close to it. Actually, it isn't even a real egg. It's a wooden one painted white. Wilma just thinks it's real. We stole the real one weeks before and put a dummy egg in its place as someone distracted Wilma with the otter's skimmer net.

The real egg is in an incubator in our office where it is safe from ravens and the elements. But we know that it's probably not fertile. Male eagles have to be fully flighted to mate successfully because they must be able to balance. Our eagles have full or partial wing amputations due to injuries sustained by gunshot. With only one wing, Sam has little chance of producing offspring. Neither pair of eagles has been successful at hatching eaglets at Grandfather. But it doesn't hurt for us to try.

Wilma is not on the nest when Pam enters the habitat, and Pam warily approaches the platform to place the eagle's food, a dead quail (a type of bird) there. Suddenly, Pam hears a noise behind her and turns to see Wilma running at her from behind a rock. Pam hurries around the pond, and Wilma dashes after her. Then Pam uses her only defense. She throws a quail at the eagle and hits Wilma in the chest with it. She throws another quail, then another as she frantically runs for the fence. Once over the barrier, she bursts into laughter, finally seeing the comedy in the situation. Wilma does not think it's funny. She ruffles her feathers, then quickly eats her quail so she can get back to sitting on her wooden egg.

Before I leave for the day, I take a new seasonal employee with me to introduce him to Yonahlossee and Kodiak. Yearling bears are different from their elders. They are not reserved at feeding time and seem to have a low blood sugar problem when hungry. I show the new employee how to give them a carrot through the fence to bide some time, then burst through the gate, trying to make it to the feeding rock. Yonahlossee has inhaled her carrot, though, and I can hear her one hundred pounds running toward me full bore, crying at every step. Before I can separate the food into two different piles, she tears the bucket from my grip. Dog food scatters in all directions.

"You'd better hurry, Kodiak!" I yell. "She's going to eat everything if you don't!"

The new employee stands wide-eyed at the gate. His expression reads, "I'm going to have to do that?"

I smile, but before I can answer, "Yes", Pam calls me on the radio. The otters have a hummingbird feeder.

A hummingbird feeder?

When I reach the otter habitat, Pam has already put the otters up and is in the habitat picking up broken glass from what is left of the hummingbird feeder. We don't find all the glass so the otter pond will have to be drained. The woman who dropped the feeder into the habitat is long gone. She has fled in embarrassment. She had been dangling the feeder over the viewing glass to get the otters' attention, and the feeder slipped from her hand. She got the otters' attention, all right. They were having such a good time playing with the broken hummingbird feeder that Pam couldn't get them to come inside for almost ten minutes after she was informed of the incident.

After cleaning up all the glass we can find and before anything else happens, I decide to make a break for it, leaving the habitats in charge of the closing shift. Tomorrow, I know, will come early. Even though I will see the same animals at morning check, I know that tomorrow will be totally different. It makes the job and life a lot more interesting. And that's the animal caretaker's life.

CHAPTER 3

NOLA THE MAGICIAN

The otters had doubled from two to four in January of 1997 and Manteo and Oconee, the first otters of the habitats, were very pleased with their new playful pals.

The visitors were thrilled, also. Four meant twice the entertainment. But the two new otters still did not have names. A public contest was going on as May approached, and the winning names were being narrowed down.

A week or so before the announcement, the new female otter we were calling "Little Girl" disappeared. I realized she was gone because one of the visitors, once again, dropped something into the exhibit. Running up to me, the woman exclaimed very apologetically, "My camera lens fell into the otter habitat, and they're running around with it!"

I ran to the otter house to get the otters inside. The four stalls of the otter house where they sleep at night each have a guillotine door that opens onto display. By sliding the doors up and down with wire cables, one can usually entice the otters inside during the day. They just can't resist the clanking of the metal doors because otters are the most curious animals.

Manteo came up to the door and peered at me, then slowly started down the steps into the dens. Next was Oconee, and then the new male otter. But the new little female did not come in.

I closed all three otters inside, then walked to another guillotine door that opened onto display and opened and closed that one ... nothing.

I walked back outside and to the upper viewing area and glanced around the exhibit. But nary a little otter body could be found. That was strange.

Maybe she's still in the water, I thought. Sometimes if an otter is underwater, it can't hear the clanking of the metal door.

Not wanting to take any chances, I climbed over the viewing area wall and dropped down into the exhibit instead of opening the door onto display. Then I covered every inch of the habitat. I looked in the water, under the log, everywhere I thought a little otter might hide. Once again, nothing. I called Kirsten over the radio, and she joined in the search. We slowly came to the realization that Little Girl was not in that habitat.

"Maybe she came inside with the others, and I just didn't see her," I said hopefully. With Kirsten following, I opened the otter house door, and we went inside. Three weasel-like faces stared up at me. Only three.

"Oh, no!" I moaned. "What are we going to do?"

"I think we need to call Harris," Kirsten instructed. Harris Prevost was the vice-president of the mountain. Slowly, I nodded. That was the last thing I wanted to do. And I felt very responsible for the incident. That morning, we had cleaned the otter pond, and I had forgotten to check on the water as it was filling back up. It had gotten so full that we had to drain some of the water out. What if the little female otter had gotten out when the water had been at its zenith? It would have been my fault.

When I called the office, Harris wasn't in. And I heard myself ask the secretary if Mr. Morton was in.

"Yes ... " the secretary said. "I'm putting you on hold."

Mr. Morton was the president and owner of Grandfather Mountain and those otters were his pride and joy. I could not believe I was going to have to tell him one of his otters had escaped. What would he say?

"Hello?" said Mr. Morton.

I started to sweat. "Huh ... Hi, Mr. Morton. How are you doing today?"

"Fine. How are you?"

"Well ... I'm not doing so good. One of the otters is gone."

"GONE!"

"Yes. Mr. Morton, I am so sorry. I don't know what in the world happened. We've looked all over that habitat and can't find her anywhere."

"Which one?"

"The little female without a name."

He was silent for a moment. Then he asked. "What about the others?"

"The rest are all there. We've put them back on display. But we can't figure out where the missing one got out."

"Well, get a net and search the area for her," Mr. Morton said. "She couldn't have gotten far."

"Yes, sir." I hung up the phone. I could have told him that a net wasn't going to do any good. If she had escaped, we were never going to see her again. But I didn't want to put any more of a damper on Mr. Morton's spirits.

So Kirsten and I, each with large eagle nets in hand that were totally inappropriate for catching otters, started around the habitat trails, then went down into the woods. We tripped over rocks and got those cumbersome nets caught in rhododendron branches, but we forged ahead after the elusive little female otter.

Steve, the trails manager, came on over the radio. He had heard the news. "Where are your other otters?"

"They're out on display," I said.

"You might want to put them back up. If they smell where she went out, they might follow."

"Oh, Lord," I said to Kirsten. "I hadn't thought of that! That's all we need is for another otter to escape!"

We came out of the woods below the bear habitats and ran back to the otter house. When we opened up one of the guillotine doors, only the unnamed male otter trotted in. We opened another door. Nothing. Manteo and Oconee were now gone!

Then we opened up another guillotine door and got the shock of our life. Hanging from the cave's ceiling just past the door was the tail end of an otter. Its tail and hind legs were sticking down. The rest of its body was inside the artificial rockwork. And to top it all off, the otter was stuck! It couldn't go forward, and it couldn't go backward.

Kirsten and I stared at the dangling otter trying to figure out what was going on. When the otter habitat had been built, the designer had used concrete to seal the little caves that the otters had to run through to get onto display. But he had forgotten to concrete the second cave's ceiling. Only sturdy metal mesh kept the otters from crawling inside the rockwork. Otters are extremely powerful. Pound for pound, they are one of the strongest mammals on Earth as well as one of the most curious. When that combination occurs, one gets into all sorts of trouble. Those little otters had chewed their way through that tough metal mesh and had peeled it back into a perfect little hole. And now one was hanging from that hole, stuck.

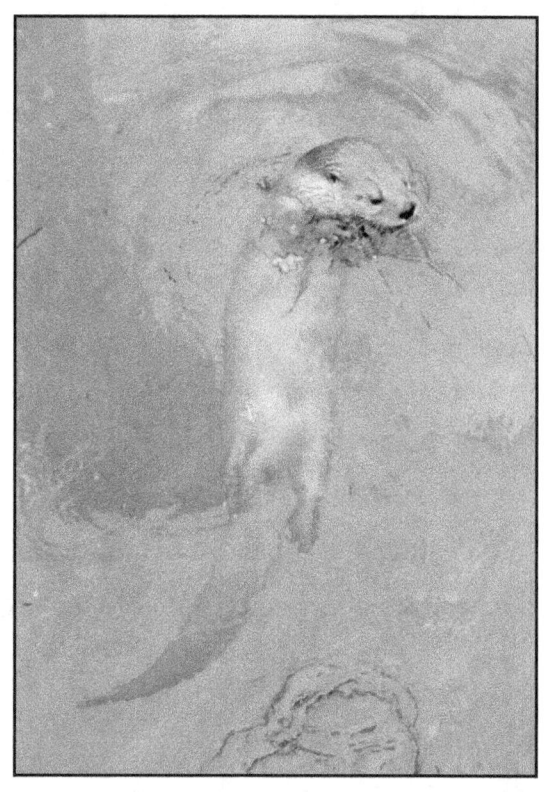

Kirsten walked into the stall and reached her hand up to the otter's wet slippery tail and gave a tug. The otter, which we thought might be Oconee, responded by thrashing its feet. but Kirsten couldn't pull it out. Then she pushed on the otter, and it disappeared into the hole. Almost immediately, Oconee and Manteo tumbled out of the hole and into the stall, and we closed the door so they couldn't get back to it.

Adorable Nola

The little unnamed female otter didn't appear from her secret hiding place, however. With relief, I called Mr. Morton and told him the news. Immediately the handyman, Harry, and his assistant, Kathleen, were called to stop up that entrance so the other three otters could still go out onto display, but not get back into that hole.

The waiting game for the missing female otter began. We kept the guillotine door open so when she came out of the hole, she could come into the inside stall, but would not be able to go out onto display.

She had not appeared before closing time, so we left a small piece of meat in that stall to lure her inside and left for the night.

In the morning, the meat was gone, but so was the otter. She had crawled back into that rockwork. The next night, we placed another piece of meat in the den and the next morning, it was gone. So was she. Because we'd only placed a tiny amount of food out for her each night, she finally got hungry. The next morning of her disappearance, there she was, curled up on the floor, asleep. Immediately, I closed the door so she couldn't get back to that hole.

The habitat designer was contacted now that Little Girl was free from her handiwork. He replaced the metal mesh, then cemented the ceiling like the rest, grumbling the whole time about how he hated otters.

We were all happy the little female had decided to return from her magical hiding place. Houdini couldn't have pulled such a performance. If it hadn't been for Manteo and Oconee, we still might not know where she went!

The week after she reappeared, she also got a name. She and the unnamed male became Nola and Chucky after the Nolichucky River.

Nola entertaining guests once more after her Houdini-like disappearance

CHAPTER 4

MR. TRASHCAN

Driving up the winding road on Grandfather Mountain one morning, I saw that every trashcan along the roadside had been knocked over ... again. And not only were the cans knocked over, trash was strewn in all directions. That annoying wild bear was back!

I dreaded walking through the habitat trails that morning because I knew every trashcan would be knocked over there as well. When "Mr. Trashcan" visited the mountain, he didn't just meander along the road. He also invited himself into the habitats.

Sure enough, as Pam, Rene', and I rounded the corner, one of the green trash barrels was turned over in front of the bald eagle habitat. Wilma, as always, spotted me. Her wild display of head feathers was sticking up crazily, and she and Sam had petrified looks on their faces. They'd been chattering even before they spotted us.

The trashcans were turned over at the bear habitat and deer habitats as well. The deer themselves looked frightened and winded, as if they'd been running. All heads were staring toward the back of their habitat and all ears were pricked in that direction. Their white tails were sticking straight up ... a sign of danger. Frowning, I walked into the deer habitat and down the hill in the direction the deer were looking. I heard a strange metallic noise.

"Pam! I think he's still here!" I called as I ran to the back of the habitat and up the large rock we called "the birthing rock" where I could get a good view of the service road.

The deer were frightened of the wild bear in their habitat!

There he was! A half-grown skinny black bear had somehow trapped himself on the service road. The ten-foot high fence and slanted chain-length arms at the top of the fence kept him from climbing out, but that hadn't kept him from trying. When he saw me, he really flipped out.

Pam reached my side and we both stared at the bear, reckoning that he'd smelled the deer's sweet feed and corn and had climbed over the rocks to get to it. When he'd heard us coming down the trail, he had run to the back of the habitat and scaled the first fence, but couldn't get over the perimeter fence because of the overhang at the top. Nothing like this had ever happened before, and we didn't know what to do.

We did know that we couldn't let that bear get away. I called maintenance for backup. Maintenance did not know what to do, either. Removing wild bears was not one of their fortes--only picking up after them was.

I knew I needed to tranquilize that bear with a dart gun so the wildlife people could take him away and release him. I was afraid that if we let him escape, he would come in after the food again. Black bears are mostly vegetarians, but the does had been dropping fawns, and I knew a bear would not pass up the opportunity of a tasty fawn snack if he stumbled across one.

We decided to move the bear down the service road to get him away from the frightened deer. The perimeter fence ran the length of the backside of the habitats, so we fixed our bear trap at the far end, past the cougar habitat, and began moving the bear in that direction. It didn't take much persuasion when he saw us. All he wanted to do was get out of our sight. He was petrified. When he ran through the gate on the service road that separated the deer habitats from the rest, we closed it so he couldn't get back to the deer.

When I called the wildlife commission to get permission to dart the bear, the "bear expert" was out, and I had to leave a message. I called the conservation officers of the county to get their permission but they didn't have the authority. So I had to wait around for a call from Mr. Bear Expert.

Finally, the man called. He wasn't much help. He said that he would have to get permission from Raleigh before I could dart the bear. So we sat back and waited.

I'd made a mistake by calling the conservation officers. Their curiosity had gotten the better of them, and they showed up at the habitat office wanting to see "Mr. Trashcan".

Against my better judgment, I agreed. I was curious myself to see if he'd been caught in that bear trap filled with sardines and cupcakes.

Why we thought the stressed out, cornered bear would think of stopping to eat cupcakes and sardines, I don't know. But I was a little disappointed when we looked down the hill and saw the cage standing empty. The bear was nowhere to be seen.

We walked along the service road, up the hill past Elizabeth's and Punkin's habitat, and when we came around the corner, there he was, pacing furiously up and down the back of the cougar habitat! When he turned and saw us, he ran toward the closed deer gate.

As we all stared with our mouths open, he scaled the ten-foot cougar fence as easily as if it was a tree and leapt atop the deer barn. In a fleeting moment of trepidation, I thought he was going into the cougar habitat, but he turned, and jumped back into the deer habitat.

Once again, the distressed deer began running for their lives as the bear ran for his, across the length of the deer habitat, onto the "birthing rock", over the back fence, and onto the service road. Again, he was trapped by that darn overhang!

I called Pam and Rene' over the radio, and we had to start the process all over again by moving the acrobatic bear back through the deer gate to the other side of the service road to get him away from the deer. The apologetic conservation officers helped us as much as they could.

Finally, in the late afternoon, I got a call from Raleigh saying I could dart and contain that crazy bear. John, the maintenance manager, the two conservation officers and Pam came onto the service road and began pushing the bear along the road past the off-exhibit bear habitats toward Rene' and me, waiting with the dart gun.

The pistol was ready, and I called to my cohorts over the radio to see if they were still headed in my direction. They were, the befuddled bear wandering and panting ahead of them. Then I saw the bear turn the corner. He was coming. My heart beat faster even though a chain link fence separated Rene' and me from the bear. My hands shook as I pulled the pistol up and positioned it through one of the holes in the fence.

Mr. Trashcan is apprehended and tranquilized

He moved up the hill, and I realized I was going to have to shoot him on the move. I had darted some of the deer previously to transport them, but this was totally different. This was a wild bear pacing frantically along with people standing in the service road behind him. What if he turned, ran down the hill, and attacked his tormentors? I worried as I aimed the pistol and pulled the trigger.

The bear gave a loud roar, swung around, and ran down the hill with the pink tuft of the dart blowing in the wind, still too frightened to get near those people at the far end. I had shot the tranquilizer high on his back. It was a good shot.

"I got him!" I yelled a little too enthusiastically. They would not admit to it, but I could tell the men in the group were definitely impressed.

The bear went under quickly, and with the help of everyone, we loaded him into the bear trap. We gave him a pan of water in case he was thirsty when he woke up and left him in the shade of a tree to await the wildlife commission's bear expert. He would have to take the bear away to find another home.

Finally, the man came to take "Mr. Trashcan" away. But our uninvited guest was not going to go quietly. When we drove the truck down for the men to load the bear cage, the bear threw a fit. He demolished his metal pan of water and dared anyone to get near that cage. He roared, showed his fearsome canines, and flailed his claws. In the end, we had to ease metal poles through his cage so we could pick it up and slide it onto the truck.

But that sly bear had one more trick up his furry sleeve! As Pam slid her pole through, the bear slapped the pole with such tremendous force that it flew backwards, catching Pam in the shin. The resulting welt sent her to the doctor several days later. A joke sprang up on the mountain about how Pam was attacked by a bar welding bear.

We all breathed a sigh of relief as the wildlife officials drove away with the bear. But "Mr. Trashcan's" relief would not come until hours later, when he was released far away from Grandfather Mountain.

Two days later as I was driving up Grandfather Mountain to work, I noticed a trash can had been knocked over. Then another. When I saw John, he stated, "That bear's back."

But if it was Mr. Trashcan, he had definitely learned his lesson. The bear had steered clear of the seemingly tranquil place where sweet feed was plentiful. In his thoughts, undoubtedly, were nightmares of dart guns, scary humans, cages, and truck rides.

The deer are safe once more!

CHAPTER 5

PHOTO BY L. L. MITCHELL

L L COOL J

If it had not been for Mr. Trashcan, a tiny fawn would never have gotten such an unlikely name as L L Cool J!

The day we found that wild bear, "Mr. Trashcan", on the service road behind the deer habitat, one of the fawns turned up missing. We searched intermittently that first day for Critter's fawn, but could not find him. At first, we were not too concerned because sometimes, no matter how hard one looks, it's almost impossible to find those little camouflaged tykes. After searching several times in the front and back deer habitats with no luck, we began to think that maybe the bear had eaten the fawn, leaving no remains.

Every day, we searched. So did Critter. The poor mother trotted all around the habitat for days grunting her call for her tiny spotted baby.

Then one day, there he was, standing on the other side of the fence in the back deer habitat and peering into the upper habitat where the rest of the herd awaited. In the melee, he must have bedded down in the large beds of jewelweed, remaining silent and still as we approached in our earnest daily searches. He had become so hungry from lack of mother's milk that he'd finally shown himself. It wasn't long before we had mother and baby together once more. The fawn drank and drank, his little tail twitching merrily as he nursed.

But all was not right with the fawn. When the wild bear had entered the habitat days before, the deer had begun running, and this little fawn must have run up onto the birthing rock and fallen into the back deer habitat, scraping his face and injuring one of his hind feet. His left jaw and right hind foot were alarmingly swollen. Because of this harrowing incident, his name began to emerge ... Lock Jaw.

Critter   photo by Hugh Morton

Lock Jaw's swollen face just would not heal, and neither would his foot. The vet gave us antibiotics to give to Critter so the baby would get some in the milk. Critter did not like the antibiotic, though, no matter how we mixed it with sweet feed, strawberry cupcakes, molasses, and jewelweed. She just wouldn't eat enough to help poor little Lock Jaw.

As the fawn's foot and jaw got worse, we realized we were going to have to catch him to treat him, which would be pure torment for us, Lock Jaw, and Critter. First we would corner him and catch him. When he cried in terror, Critter would come running, grunting her concerns. Once, when we caught him in the barn to treat him, Critter ran over a hay bale and tripped all over it in an attempt to make sure her baby was all right. She would always hover around until we were finished treating him, then lick his fears away when we finally let him go.

Slowly, after weeks of distress, Lock Jaw's wounds began to heal. His name began to go through a metamorphosis, shortened to L. J. and later jokingly expanded to L L Cool J after the famous rap singer. The name stuck.

L L Cool J survived the trauma of the bear episode and the treating of his injuries, but all these fawnhood disturbances left him somewhat stunted in growth. Long after the other does had stopped nursing their fawns, Critter continued to nurse Cool J, even though he was capable of eating the sweet feed and deer chow provided.

Then came probably the most traumatic event in poor little L L Cool J's life. In late spring, before the new fawns are born, the habitat staff releases the yearling deer by herding them onto the service road and throwing open the gates. They step into an unfenced world where they lead full, sheltered lives in the 5,000 acre nature park, free from hunters. They join the small herds of deer that visitors see scampering across the road in the early mornings or late afternoons.

We had released the yearling deer for several seasons this way without any problem. There was a gate on the service road directly behind the deer habitat, but I didn't like to use that one. I was afraid that if we released them that close to the deer habitat, the yearlings would not want to leave the area. In the past, some had actually jumped back in. By driving them down the service road the other way, behind the cougar and off-exhibit bear habitats, they could leave the confines of the enclosures several acres away. When they did flee through the gate, they would always run in the opposite direction from the deer habitat.

The first day, we released Heidi's and Star's little bucks. We had tried to get L L Cool J to follow them, but he would have none of that! He wanted to stay in the habitats with his mother.

The two yearling bucks ran down the service road, nervously sniffing at the strange smells of cougar and bear. They would stop and look back at us, then run farther down the road. Finally, when they reached the open gate, they bounded through and ran off together into the woods, just as expected.

The next day, we tried once more to get L L Cool J out onto the service road. After thirty minutes of chasing him around the display, we finally got him to run out and started the process of pressing him down the service road.

Everything was going well at first. The little deer passed the cougar fences and continued to walk down the road hesitantly past Elizabeth's and Punkin's habitat. But when he came to the yearling bear habitat, things began to go awry.

Carolina and Dakota, who were in the yearling habitat that year, spotted him from a high rock and were instantly curious. Slowly, they started to meander down to see the strange looking brown animal with two spiked antlers sticking out of its head.

Cool J didn't see them at first, but I tried to speed up the procedure just the same. He was almost there ... to the open gate and freedom.

Suddenly, he stopped and looked to his left. Two frightening black furry creatures were moving his way! He turned around and looked at us uncertainly.

"If he comes this way, stick your arms out and form a line," I said to Rene' and Tanya. "I'm sure he'll keep moving."

At that moment, he decided to take his chances with us. He started running back to the safety of where he'd come.

"Don't let him pass!" I yelled.

He was running right at Rene', and they played "chicken" until the last possible second. Rene' dove out of the way, and Cool J ran up the hill in the direction of the deer habitat. But we had closed that part of the service road, so he couldn't get any farther than the back of the cougar habitat. Irritated, we tried again.

Bear Pranksters Dakota & Carolina   photo by L. L. Mitchell

We walked up the hill and started him back down again. This time, Carolina and Dakota were ready, thinking this was a hilarious game, and as Cool J came running down the service road, they jumped up on the fence at him.

The little deer turned, once again intent on plowing through us, and this time played "chicken" with me. He jumped clean over our heads and ran up the service road to the back of the cougar habitat again.

By this time, we realized releasing Cool J this way was not going to work. As we trudged up the road, I decided to put him back in with the other deer for the night and try something different the next day.

When we rounded the corner, we saw that Cool J had now gotten the cougars' attention. He was standing next to the cougars' habitat and I could see, through the tangle of blackberry vines, a long, tawny colored tail running down the fence toward the deer.

"Roooow!" went Sheaba the cougar.

Cool J ran along the fence and into the blackberry bushes with all three cougars in hot pursuit. The chain-link fence separated them, but poor Cool J didn't know that. Suddenly, Cool J burst from the vines and galloped down the service road toward Carolina and Dakota.

"Get the gate open!" I yelled at Tanya, so Cool J could run back toward the other deer. Before she could move, we heard the galloping of little deer hooves, and here he came, reeling around the corner. He stopped at the closed gate with tongue lolling and terror-stricken eyes, then turned and galloped down the road again toward Carolina and Dakota.

The next time he came galloping by, we had opened the gate, and he ran down the service road toward the deer. Now that we had him on the deer side of the service road, we couldn't get him to go back into the deer habitat. He ran up and down the fence for what seemed like hours before we finally got him back in.

To make matters worse, the other deer seemed to have forgotten who he was. All of them chased him around for a long while before everyone finally settled down. I was just relieved that he had not had a heart attack.

That part of the service road where Cool J had galloped up and down, darting around cougars and bears, became forever known as The Bear and Cougar Gauntlet. Poor Cool J had failed miserably.

The Bear & Cougar Gauntlet

The next day, we decided to let Cool J out through the open gates on the deer side of the service road even though I was afraid he might try to jump back into the deer habitat. This time, everything went like clockwork. The small deer ran into the woods, and we thought we'd seen the last of L L Cool J.

Two months later, Tanya and Rene' were counting the deer at morning check. Tanya noticed our buck, The Count, standing beside the pond and saw several of the does meandering around.

She didn't see all the deer from her vantage point, so she started down toward the gate to count the rest. Standing near the gate was The Count. She frowned. How could The Count have gotten from the front of the habitat to the back in that short of time? She looked closer. It wasn't The Count. It was L L Cool J!

After two months of living in the wilderness, he'd decided to jump back into the deer habitat. But the other deer had not taken kindly to Cool J's return. They obviously had been chasing him around all night, kicking at him. He had scrapes all over. As Tanya watched in dismay, several of the does appeared and began chasing him again.

Cool J was so tired and miserable from being chased around all night by his former companions that he let himself be herded easily and ran off without tarrying when the gates were thrown open for him.

And that really was the last time we saw Cool J. Even if he had found his way back to the deer habitats again, he probably would have thought twice about jumping in once more after his unpopular return. Like "Mr. Trashcan", he decided to stay away from the place where sweet feed was plentiful.

Dakota, Tanya and the Habitat Truck   Photo by L. L. Mitchell

CHAPTER 6

DAKOTA'S WILD RIDE

"Habitats?" Martha the peanut lady said over the radio. "Did you put Carolina and Dakota in the big bear habitat?"

Tanya and I looked at one another and frowned. "No," I answered. "They're still in the yearling habitat."

There was silence from the other end. Then Martha said, "I think you need to come down to the bear hut."

Both of us hurried down the path toward the bear hut, the little shack in front of the big bear habitat where guests can buy peanuts and apples to feed the bears and deer. Martha was standing by the rail looking into the habitat.

Sure enough, sitting right up front begging for peanuts were Carolina and Dakota. Mumbles, our large male bear, was sitting beside them, never missing a beat in catching his peanuts. He didn't seem to care that they were begging beside him.

We stood there in shock for a moment. How did Carolina and Dakota get in there? Those two were still too young to be in with the big bears. But as we continued to watch them, we realized they were getting along fine with Mumbles. He obviously was not going to do anything to them. It looked so funny, the large six-hundred pound boar and two tiny yearling sows sitting beside him, all catching peanuts.

Martha was happy. She had complained several days before that the bears were all "dead". They weren't, of course, but some of our adult bears had gotten so old and arthritic that they slept in the shade or behind rocks where visitors couldn't see them instead of moving around and catching peanuts. Now, with two yearling bears up front, there was a lot more action, and the visitors loved it.

Suddenly, Gerry, one of the adult females, appeared from behind the rocks at the pond and spotted the two luckless yearlings. She darted toward them, and they jumped up and ran out of view. Dakota and Carolina were neck and neck as they came around the corner again, trying to distance themselves from Gerry. As they passed in front of Mumbles, he watched them, perplexed, then went back to catching peanuts.

We realized that leaving Gerry with Carolina and Dakota would not be a good idea. Gerry was just being a normal territorial bear, but there was no way for them to get away from her in the two-acre habitat. It didn't seem like she was going to relinquish her reign anytime soon. We would have to get Carolina and Dakota out of there.

We entered the habitat to stop the game of chase, but when Carolina and Dakota saw us standing there, they tried to hide behind us! We had to jump out of the way as Gerry ran toward us to get to them. Gerry then ran out of view. After some discussion, we decided to move Gerry into the yearling habitat instead of putting Carolina and Dakota back there. The public had greatly enjoyed the two younger bears' antics. If they continued to get along with Mumbles and with Maxi, who was too old to chase them around, they would be allowed to stay up front. We also found out how they'd managed to sneak up front. They'd dug a huge hole that would have to be cemented!

Because of the size difference between Mumbles and his two companions, however, visitors began to think that Mumbles was a mother bear and Carolina and Dakota were his cubs. It was funny to us keepers as we sat back and listened to visitors' remarks.

Martha didn't think it was all that funny. It was hurting her peanut sales. There was a sign on the peanut and apple box that read, "Please do not feed the mother bear and cubs", because visitors were not allowed to feed the *real* mother bear and cubs in the next habitat. Since many visitors thought that Mumbles was the mother and Carolina and Dakota were the cubs, peanut sales dropped off, much to Martha's and the big bears' consternation.

Things went fine at first besides visitors confusing who Mumbles, Carolina and Dakota were. Mumbles acted as if the two yearlings had always been there, and Carolina and Dakota learned to catch peanuts like pros (when given the

opportunity). Carolina had a cute habit of sitting up and holding her right hind foot with her front paw as she scanned the crowd for potential subjects. Dakota, however, could not leave well enough alone. For some reason, she seemed to have a death wish, and began to torment Mumbles.

We noticed her brazenness for the first time one morning when we went in the feed the bears. I gave Mumbles his apples, carrots and other vegetables and dog food, then led Dakota to her rock to feed her. Dakota wasn't content with her food, though; she wanted what Mumbles had.

Slowly, she sneaked up behind him, and when he turned his head in the other direction, she grabbed his remaining apple and ran off with it. He realized what was going on and snorted at her, but went back to eating his food.

Dakota sneaked up behind Mumbles after she'd eaten his apple and this time managed to steal one of his carrots. Mumbles jumped up and chased her this time, but his efforts were only half-hearted. Grumbling, he settled back down with his food.

Later that day as I passed the big bear habitat, I glanced in to see if everything was going well. There were Mumbles and Carolina as usual sitting up for peanuts. Dakota was creeping slowly behind Mumbles with a mischievous look on her face.

All of a sudden, she galloped up to him and swatted him hard on the rear end, then wheeled around to run. Mumbles growled as he turned around and chased her out of view. It wasn't long before he was back up front catching peanuts, and Dakota was back up front tormenting him. It was comical in a way, but we all wondered when the day would come that Dakota would get her due.

Several days later, Dakota was up to her tricks again, and Mumbles had the most exasperated look on his face. Already that morning, she had stolen his lettuce and swatted him on the back when he wasn't looking. As the afternoon had progressed, she had impeded his peanut catching tactics. As she ran up behind him to swat him for the third time that day, Mumbles had finally had enough. His patience, which had been a virtue, was gone.

Dakota realized the difference when Mumbles turned around to charge after her. He meant business. She fled behind the pond with the mammoth bear barreling down on her and tripped as she ran down the steep hill.

From the overlook, we heard Dakota cry out in pain. Silence ensued. Mumbles ambled from behind the pond and sat down to catch peanuts once more. Dakota didn't come back. Tanya and I entered the bear habitat to find her. When she heard us coming, she made her way to us, moaning. She was walking on three legs and holding her left front leg up. She wouldn't let it touch the ground.

"It serves you right, Dakota," I said. "Next time, you'll think twice about messing with Mumbles."

But Mumbles had not hurt her. She had hurt herself. In her haste to get away from him, she had fallen down the hill and twisted her elbow. We moved her into "the neck", the bear infirmary that is off-exhibit, and called Howard the vet.

Howard suggested we watch her for a couple of days. If she did not get better, we would have to take x-rays to see if she had broken anything. She did not improve, so Howard made arrangements at Sloop Hospital in Crossnore about ten miles from Grandfather and came the next day to sedate Dakota so we could take her for x-rays. After Howard darted her, we all sat back and waited for her to fall asleep.

"How are we going to get her to the hospital?" Pam asked.

"I guess on the truck," I said.

Pam was silent for a moment. Then she said, "You mean you aren't going to put her in a cage or anything?"

"We don't need a cage," Howard said. "If she starts coming around, I can just inject her with some more tranquilizer."

Pam didn't look convinced, but she didn't say anything else about it.

When Dakota finally decided to go visit the sandman, we opened the gate to "the neck" and all four of us pushed and pulled her onto the back of the truck. We drove the truck up to the habitat office to call the hospital and let them know we were coming.

Again, Pam piped up. "Are you really going to take her to the hospital like that?"

Tanya, seeing Pam's concern, tried to make her feel better. She went into the office and grabbed a large leather collar and a chain, then came back out and placed it next to Dakota in the back of the truck. "If she starts coming to, I'll just put that collar around her neck and hold onto the chain," Tanya said.

After Howard had made his phone call to the hospital, he came back outside and climbed into the truck. I would drive, Howard would ride shotgun, and Tanya would ride with Dakota. Howard gave Tanya a syringe of tranquilizer just in case Dakota woke up.

Pam came out of the office ashen-faced and very serious looking, holding the largest dog carrier that we possessed. "Maybe you could take this in case she wakes up," she said.

I smiled as I took it, imagining us trying to get a 120-pound half-awake bear into a carrier made for a 50-pound dog. But it made Pam feel better.

With everyone in position, we made a dash for the hospital. Tanya had to hold on to Dakota as we took the winding mountain curves. I kept my eye on the road while Howard watched out the back window to make sure everything was all right.

As we barreled down the highway, Howard suddenly stated, "Looks like she might be waking up a little. Maybe you should pull over."

"Oh, Lord!" I thought aloud as I pulled the truck off the road.

We jumped out of the truck and into the back with Tanya. Tanya whipped out the syringe and injected the groggy bear. We were all very relieved when Dakota went back to dreaming.

Again, I pulled back onto the road, arriving at the hospital ten minutes later. Dakota wasn't actually allowed inside the hospital. Howard had brought along a portable x-ray machine so he could take x-rays in the truck, then develop the film inside the hospital. We found a suitable shady spot behind the post office to park the truck.

We let the tailgate down and pulled Dakota onto it. Howard pointed the x-ray machine at her elbow and got several pictures. He was gone a while and Dakota began to wake up again. But Tanya gave her another shot, and she went back to sleep.

Howard came back out to get a couple more x-rays and discovered that Dakota had begun to draw a small crowd. It isn't every day that a person venturing to the hospital sees a sedated bear getting x-rays in the parking lot!

After viewing the x-rays, Howard determined Dakota had not broken her elbow. She just had a bad sprain and time would eventually heal it. We were all happy with the news.

We closed the tailgate and began the speedy ride back to the mountain, this time with no problems. When Dakota woke up again, she was back in "the neck", where she had to stay for several weeks until her elbow healed. When she was finally moved back to the big bear habitat, she occasionally played pranks on Mumbles, but not like she had before, and she was much more wary when she did.

Only Dakota knows if she remembers fleeting glimpses of the wild ride in the habitat truck, but it is a memory that her keepers will never forget.

Dakota "fishing" for lost peanuts in the Big Bear Pond

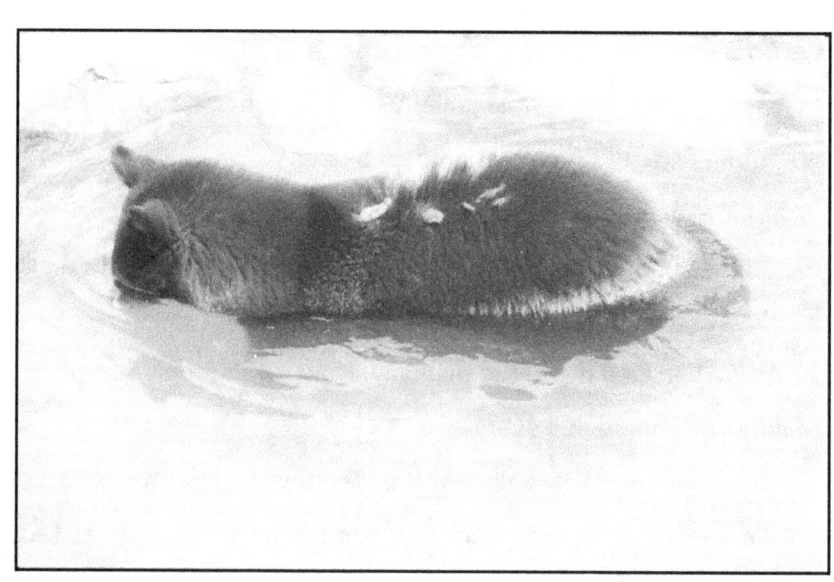

Photo by L. L. Mitchell

Maxi the Bear   Photos by L. L. Mitchell

CHAPTER 7

VET CALL

All the animals were fine at morning check, and we are trying to get all the morning chores done before the vet arrives. Today, we have scheduled rabies vaccinations, and the vet is coming to try out our new dart gun.

Howard the vet and Tanya like this new contraption, but I don't like change all that well. I like my little dart pistol. We've been through a lot together. Even I have to admit this new dart gun is state-of-the-art. It's practically silent when fired and can be set at different gauge settings for short or long distances. Setting it is the tricky part. Howard has taken the gun home to test it on his neighbor's unsuspecting cow to make sure at what number to set the pressure gauge.

Thank goodness he practiced on a wall of his home first. He set the pressure gauge on six, fired, and put a hole through the wall! For the cow, he had tried a conservative two setting and that seemed to do the trick. The cow had jumped a little as the dart had hit her, but had gone back to eating grass. It never noticed the pink tail of the dart sticking out of its flank or saw it drop to the ground moments later. Feeling confident with the new dart gun, the vet has put our animals on his next agenda.

The vet has not yet arrived, so Tanya and I go about the daily tasks. After cleaning out the cougar house, we walk back up to the otter's viewing room to turn on the deer waterfall. The switch is in a cabinet inside the room, and for the past two weeks, we've tried to be careful when we open it. A mother deer mouse has decided that the cabinet is a good place to raise a family, and we don't want any mice falling out.

Many people might think we're crazy, but we think those little mice are adorable and don't have the hearts to get rid of them or their nest.

This morning, Tanya gives them a little help. As she opens the cabinet, two of them fall out, hit the floor, and take off on their little mouse feet in different directions. One runs into the otter viewing room and the other heads for the bright daylight at the exit. Another falls, then another, until there are about five or six fuzzy gray mouslings running helter skelter for cover.

A young girl, unaware of the intimidating mice just inside the otter viewing room, turns the corner and sees three running right at her. She screams and flees to the safety of the outdoors. In her haste, she trips and falls. As the mice flee the room past the dazed girl and into daylight, a toddling three-year-old boy spots them. In his excitement, he begins chasing a petrified baby mouse, but loses it when it runs up a drainpipe. The boy's parents laugh. He's been throughout the whole habitats and has seen the bears, cougars, otters, eagles, and deer, and his favorite thing has been the baby mice.

Baby mice and chipmunks, I think. Why do people get so excited about baby mice and chipmunks? Finally, all the little mice have escaped, and the crowd scatters to go look at the deer and other creatures. It gives us a chance to sneak away to feed the bears.

When we enter the bear habitat, Dakota is jumping up and down, trying to catch a butterfly. When she sees us standing at the gate with food, she heads our way.

Maxi, our oldest bear, is feeling arthritic, so we don't make her walk to the front of the habitat like we normally do. She needs a pedicure desperately, and we make a mental note to tell the vet when he arrives. Twice a year, Maxi has to be tranquilized so the vet can trim her claws with wire cutters. She is so old that she can't move around like the younger bears, and therefore doesn't wear her claws down. They have begun to grow too long, impeding her walking ability.

"Habitats?" It's the gate calling. "The vet is on his way up." We leave the bear habitat and walk back up the hill to the office. Howard is already there, pulling that dart gun out of his truck.

"Ready?" he grins. We go to the deer habitat first and Howard gets the dart gun ready while Tanya acts as his assistant. It doesn't take him long to load the rabies vaccine.

The Count is the first deer patient. Howard walks in, aims and shoots. At the initial shot, The Count jumps, then looks back at his flank. When he sees the pink fuzzy thing sticking out of his rump, his eyes widen and his expressions says, "Oh my gosh! What's that?!" He takes off at a gallop.

The dart does not fall out like it's supposed to, much to The Count's consternation. He stops, looks back at the dart, and takes off again, trying to run away from that dreadful pink thing. Again, he stops. He looks back, sees the pink thing, and runs. Around and around the habitat he goes until ten minutes later, the dart mercifully falls out. The other deer are harder to dart because The Count has stirred everyone up. Finally, they all get their shots. The bears are next.

Tanya decides she would like to try her luck with the new gun. Unsuspecting Mumbles is up front catching peanuts from visitors as Tanya climbs over the fence and aims down at him. His ears prick forward, and he stares at her as if to ask what in the world is she doing.

Moments later, he finds out. When Tanya shoots the gun, Mumbles thrashes around, jumps up and flees to the back of the habitat with the pink tail of the dart hanging out, and we don't see him again for the rest of the day.

Carolina is the next target. We give her an apple for distraction, and Howard darts her right in the hind end. She looks around and can barely see the pink dart dangling near her tail. She tries to reach it with her front paw. Round and round in a circle she goes, her head cocked at a funny angle to keep sight of that pink thing. Finally, Tanya moves around her and pulls it out. Carolina sniffs it and walks away toward the visitors and peanuts.

Dakota is having none of this undignified clown-like behavior. When Howard darts her, she promptly turns around, pulls the dart out, and demolishes it.

Maxi is the last one on the list, and she needs more than just a rabies shot. It's also her pedicure day. Howard decides to tranquilize her first. After she's down, he can inject her with the rabies vaccine.

No one would ever believe how fast poor old Maxi can run when she is darted with a gun. She acts as if she's miraculously cured of her arthritis. Down the hill she goes, blowing, huffing, and grumbling. After twenty minutes tick by, she finally goes under, and Howard whips out the wire cutters to trim her claws. With the work done, I am left with Maxi until she wakes up so I can defend her from those playful pranksters, Carolina and Dakota.

Maxi getting her pedicure

After Maxi recovers, I decide to get Sir Hiss, our black rat snake, out for the public. But halfway to the habitats, I stop at my car to retrieve something and place the snake on the steering wheel. What harm can he get in there? The answer is plenty. Before I realize what is happening, he has found a hole under the steering wheel and is slithering into it. I grab onto his tail and pull gently. But Sir Hiss is a constrictor, and each time I loosen my hold, he slides in and contracts his muscles anticipating the next tug. Each time I try to get a better grip, he slips further and further into the hole. Finally, he's gone ... lost forever in the depths of my dashboard. My mother, who just so happens to be visiting that day, swears she will never ride in my car again.

With the visitors beginning to pour in, Tanya and I decide to get Milton out of the closet and hand out candy to the little children. Pam will have nothing to do with Milton. He makes her claustrophobic.

Milton is a bear costume. With the head on, the wearer can hardly see anything. Another person must walk around with Milton so that he doesn't run into people, trip, fall, or hit his head on low hanging branches.

Today, I am Milton. I put the body on, then the head, and Tanya and I start out into the habitats. Tanya carries a bag of candy.

Everything goes well at first. Tanya leads me down the path to the front of the otter house. At this point, children bombard Milton. Parents have their little ones pose for pictures with the smiling bear.

Inside the costume, I'm starting to sweat and people are knocking me around. There are so many children milling around that I can't see all of them from my two small eyeholes. Tanya is somewhere to my right, and I turn my big Milton head to try and find her. What I don't realize is that a little boy is standing right beside me, petting my hand.

Whap! goes my hand as I turn around. I hit the unlucky boy right in the head. I feel the contact, and try to adjust my eyeholes to see what I've done. That's when I see the small towheaded boy, face all scrunched up to cry, hurrying back to his father. I place my hands on Milton's cheeks and whirl around in Tanya's direction.

She comes into focus. "Uh, oh ... " she says.

By this time, the little boy has reached his father and has burst into tears. Trying to rectify the situation, Tanya walks over to the crying child to give him a piece of candy. I want to rectify the situation, too. I want this boy to like Milton. So I start in his direction.

But when the boy sees the smiling bear approaching, he becomes terror-stricken and screams.

"Stay back, Milton!" Tanya instructs.

That poor little boy will probably be forever fearful of costume animals. All because of my tunnel vision! I walk away, totally dejected, and fail to see the long branch hanging over the railing.

Whap! goes Milton's head on the limb. I hit it with so much force that the limb hurls me back. I trip over my buffoon-sized tennis shoes and almost fall down. The visitors around me begin to laugh.

Tanya is suddenly back at my side. "Let's get you back to the office, Milton. I think you've done enough damage for one day!"

Next time, someone else will have to be Milton. I've humiliated myself for a month of Sundays.

Elizabeth coming out of her den to check out the snow   photos by L. L. Mitchell

CHAPTER 8

SNOWSTORM!

The weather had been deteriorating all afternoon, and the steady rain was beginning to turn to snow. The weather report predicted that the temperature would slowly rise throughout the day, but we had seen the opposite. When the snow started falling, it came down with a ferocity that I had never seen. Seasoned mountain people began nodding their heads. This definitely was going to be a big one. A Huge One.

By mid-afternoon, the habitat paths were covered with a half-foot of snow, as were the roads. Even though the maintenance crew was trying hard to keep the road up Grandfather Mountain plowed, they were losing ground. We decided to put the cougars and otters in their houses and call it a day.

The cougars were happy enough to come in. They didn't enjoy the pouring snow one bit! Watching them head toward the indoor dens was a lot of fun. At each step, they would pick up a big round foot and sling it, trying to discard the clinging white fluff. They were thankful to go to their dens for a meal, even though it was only one o'clock in the afternoon.

The otters were a different story. Otters LOVE playing in the snow. They make tunnels and slides, and when their pond freezes over, they have a great time making little holes in the ice, splashing into the water and swimming under it. They look like little seals as they come up to the "blowhole", then once again disappear under the ice.

We walked to the upper viewing area to look at them. Oconee was having a running fit and the other three were not far behind. They would gain speed, then jump and skid on their little bellies down the hill. They were fun to watch, but we were worried about getting home in the growing snowstorm.

The otters filed in through the door when we opened it, and they ran into their dens for the food we'd laid out. Their little faces showed disappointment that we were already locking them in for the night.

The deer were bedded down close together amid the trees, chewing their cud, while snow fell upon their backs. Their ears were pressed back and their heads were bent forward as if readying themselves for the storm. They have a barn, but they hardly ever use it, preferring to take their chances with the elements.

The eagles were already in their caves and the bears were asleep in their dens. It was time for us to go home.

Warily, we started down the mountain in first gear. Near the bottom, I felt my Jeep sliding. Under the mounting snow was a layer of ice. Four-wheel-drive or not, I was going into the ditch. My Jeep seemed to move in slow motion as I began sliding sideways. The car stopped just as easily as it had started into the shallow ditch. Putting it in four-wheel-drive low, I slowly eased out of the ditch and passed through the entrance gate, leaving telltale signs of a Jeep snow angel. For some reason, everyone else thought it was funny.

Tanya and I followed each other down the precarious Yonahlossee Road toward Linville. When we finally reached the narrow valley, we sighed with relief and drove home as safely as possible, passing other cars that had gotten stuck along the roadside in the soft snowy shoulder.

Once home, I made a roaring fire and sat back, watching the snowfall. As darkness took over my view of Grandfather Mountain, the snow was still coming down hard.

By morning, the snow had fallen almost three feet and was still coming down! I had never seen anything like this in my life. In disbelief, I wandered out to the parking lot where my Jeep awaited. Every car in the parking lot was buried under snow. I trudged up the steep stairs of my condo and worked my way to the front of my Jeep, staring at the mound of snow. Somewhere under there was my red Cherokee, but the snow on the ground was several inches above where the headlights should be. That Jeep wasn't going to budge.

How was I going to get to the animals? They needed to eat. Their dens needed to be cleaned. The cougars and otters had to be let out of their houses for a frolic in the snow. But I was helpless to do anything about it.

My dog, Cody, ran up behind me, almost knocking me down, and bounded into the snow-filled parking lot. The snow came up to his chest and he was having a hard time getting around, but he wasn't complaining. He picked up a mouthful of snow, ate it, and took off, bounding and jumping with the snow falling hard all around.

I worked my way back in and called Kirsten, who'd been staying with a friend who lived in a cabin on the Blue Ridge Parkway.

"Hello?" Kirsten said.

"Kirsten? This is Laurie. How is the snow over there?"

"Well, I don't think it's *that* bad," she said.

"Have you looked?" I couldn't believe it had snowed that much more at my place.

"We haven't been out yet, but it doesn't look so bad from the window."

"I've been out, Kirsten, and it's terrible. The snow is at least three feet deep. There's no way I can make it up to Grandfather to check on the animals, and the radio says there's a curfew. No one is supposed to be on the roads they're so bad."

"I'll come get you in a while," Kirsten answered matter-of-factly, as if she had complete control of the situation and could easily reach her destination.

I called the Grandfather Mountain gate repeatedly, but no one answered. I called the museum, the office. Still no answer. I kept listening to the weather reports on the radio since the television cable was out and waited for Kirsten to arrive.

Instead of Kirsten driving up to my condo, I got a call from her instead. She sounded a little sheepish.

"Kirsten, where are you?" I asked.

"I'm in Blowing Rock. Right after you called, one of the Parkway employees with a backhoe showed up to check on us. He had to use the backhoe to get us out!"

I knew Kirsten had been staying at least five miles from Blowing Rock. And I felt that if I couldn't get anyone at Grandfather, then the Yonahlossee Road from Linville (the road where Grandfather Mountain's entrance is located) was blocked.

"What are we going to do?" I asked. "How are we going to check on the animals?"

"I don't know," Kirsten said. "Maybe the snow will let up soon. But there's no need to try to make it to the mountain if no one's answering at the gate."

I got off the phone and made another cup of coffee, sat next to the window and watched the snow fall. It showed no signs of slowing down.

I called several other mountain employees' homes. Everyone was snowed in. There was no way I would be able to make it up to Grandfather to feed the animals.

I could just imagine Mina, Sheaba and Squeak staring through the bars of their dens and wondering when we were going to arrive to let them out. The otters must be going bonkers. They would be bouncing off the walls by now; they knew what white fluffy fun they were missing just beyond their little guillotine doors.

As night fell and the snow slowed to a flurry, I hoped I'd be able to reach the animals the next morning. I was sure they would be extremely happy to see me.

To my relief, in the morning the sky was cerulean blue. That did not exactly solve my problem! My Jeep was still under three feet of snow in the parking lot. The snow was too deep for the condo maintenance crew to plow using a normal truck. We learned we'd have two more days of waiting until they could get a backhoe to dig everyone out.

Thankfully, I didn't have to wait that long to escape my condo prison. Kirsten had managed to make her way to my residence because the road crews had been working all night, and she had picked up Tanya as well. We cautiously started toward Grandfather Mountain.

I was shocked by the amount of snow everywhere. The gas station at Tynecastle had really felt the effects. The roof above the gas pumps had given way under the strain, and twisted metal and snow hid the pumps from view. It would be a while before anyone used them again.

Finally, we reached the Grandfather Mountain gate. The maintenance crew was already there and had been working tirelessly on the road that wound up to the habitats and museum. We were happy we'd parked the habitat truck at the bottom so we could drive up.

Kirsten started the truck to warm it up, and we began loading our supplies. But before we could pile in, I saw John the maintenance manager approaching.

"The road is only plowed around this first turn," he said. "The drifts are so high that we have to rent a backhoe. You'uns are going to have to walk to the habitats."

All three of us stared at John with our mouths open, not believing our ears. We were going to have to struggle on foot through waist high drifts and up the winding mountain road to reach the habitats? The

animals were at least a mile up. We thought he was kidding. We *hoped* he was kidding.

As he stared back at us, it suddenly dawned on us that this was not a joke. In disbelief, we wearily pulled our packs out of the truck, looking longingly at the vehicle. Then we walked back to the newly plowed road and started our long climb.

As we made the first turn, we saw Richard in the Kubota trying his best to remove all that snow. Beyond him was a vast white no man's land. As we passed, we waved and bid everyone farewell. Immediately, we sank up to our knees in the powder.

Around the second turn, I ventured too close to the side of the road and went off into the ditch up to my waist. We were laughing so hard at my plight that Kirsten had a difficult time pulling me back out. From then on, we formed a line, each taking turns as leader to make a trail for the ones following. The going was slow, and we took many breaks. But the weather was fine, and the wind was low. We made it to the habitats over an hour later.

When we saw the snow-covered habitat paths, we cringed. It was our job to shovel from the entrance path at the museum to the bear habitat, and there were five and six foot drifts in places. Shoveling the paths was not on the agenda that day, however. Our priority was making sure the animals were all right. We headed to the habitat office to get food for everyone and to retrieve the blowtorch and shovels so that we could get the gates open.

We also whipped out our radios and called maintenance to tell them we'd made it. We were once again in contact with the outside world.

The electricity had gone out in the storm; therefore the water pump didn't work. To clean the animals' dens and provide them with water, we had to melt snow in buckets in front of the office propane heater.

Slowly, we trudged back into the habitats. The eagles were nowhere to be seen, but we could hear them chattering in their caves under the rocks.

Blowtorching open the frozen otter doors resulted in some pretty excited otters. When we opened their guillotine door, they looked out with astonishment, then all piled out. Immediately, they dove into the deep snow, making tunnels and chasing each other. They ran and ran, then slid on their bellies down to the frozen pond, where they promptly began a form of otter ice-skating. Before long, they'd broken a hole in the ice and were swimming in the freezing water beneath.

The deer were all standing side by side at the gate as if to ask, "What took you so long? We're extremely hungry and want something to eat, please!" As we passed through the cougar house to get the deer feed, we heard relieved meows and purrs from the cougars. Like all cats, though, they had to act a little standoffish at being ignored for two days.

The animals were happy to see us!

When their doors were opened, the cougars just looked back at us as if to say, "You want us to go out there? We don't *think* so!"

The deer were practically jumping up and down when we returned with their food, but we couldn't get through the gate, which opened in. The snow kept us from pushing it open. Someone would have to go over the icy deer pond rocks and dig the gate out from the other side. That's the bad thing about fence building. It's done in the summer months when no one pays attention to what might happen in three-foot snowstorms in the winter!

Finally, the deer and other animals had been checked and fed, more gates had been dug out, all fence lines checked for downed trees, and the dens washed out with slushy snow. We wearily called maintenance to see how far they'd made it up the mountain, hoping that someone could drive to pick us up. John informed us that the backhoe could not get there until the next day. We were going to have to hike back down.

Late in the afternoon, we began the journey back down the mountain. When we finally reached semi-civilization again, we saw that Richard had barely made a dent in all those drifts. With growing dread, we realized we were going to have the fun experience of trudging through all that snow once more to reach our animals if the backhoe didn't arrive in the morning.

The next morning did indeed find our exhausted bodies once again assaulting the snow-filled road on Grandfather. By mid-afternoon, the backhoe had arrived and had plowed its way to the museum. It was a good feeling to know we wouldn't have to hike down, but our work was not over. We now began shoveling all the habitat paths with the help of employees from other departments who pitched in. It took several days of back breaking work for the paths to be cleared enough for visitors to once again pass through the gates.

After that, I cringed whenever the snow began to fall. Never again would I look at it in quite the same way. Beautiful fluffy flakes gliding effortlessly and silently among leafless maples and evergreen rhododendrons to pile like soft down on branches, leaves of yesteryear, and frozen lakes appealed to me no more. All I saw was the assault on Grandfather in waist deep snow, hungry animal faces, and snow shovels.

The author exhausted after shoveling mounds of snow!

Laurie and Tanya with Yonahlossee and Kodiak   photo by Hugh Morton

CHAPTER 9

A CUB TALE

Almost everyone loves bears, from the young child who clings to her teddy bear at bedtime to the aging adult who is thrilled at seeing his first living bear in the wild. There is no question as to why people are fascinated. We all grew up with Paddington, the Berenstain Bears, and Winnie-the-Pooh. We found out through childhood stories that bears are not too different from us. There is something vaguely human about them. Bears are comical and cute, and bear cubs are downright cuddly. Cuddly? Well, maybe that *is* a bit of an overstatement.

The animal habitat staff at Grandfather Mountain had the experience of hand-raising bear cubs in 1999. We obtained the two three-month old cubs that May from Bear Country U.S.A., a drive-through animal park in South Dakota. Bear Country has the largest collection of black bears in the world - close to

200 of the bruins. Because black bears are hostile to cubs that are not their own, the animal staff at Bear Country must pull the cubs away from the sows before the cubs are old enough to leave the dens.

There is no way for the staff to isolate every mother with cubs. After pulling the crying, hungry cubs from the dens, the staff starts them on bottles, and then looks for zoos or trainers needing bear cubs.

Since none of the bears on Grandfather Mountain had cubs that year, we fit the bill. The two cubs were flown in from South Dakota, and Tanya and I met them at the Charlotte, NC airport. They were so tiny--around six pounds--but still a lot bigger than their eight-ounce birth weight. They were scared, exhausted and hungry--and that made them quite ornery. To top it all off, the employees in cargo thought they were dogs!

I had to separate them in their carrier on the way home because the female bear kept attacking the much smaller male. The little male finally ended up on my lap, where he promptly went to sleep. The female cub, however, cried incessantly the rest of the way home since her furry little punching bag was gone.

The habitat staff had tried to think of just the right names before the cubs' arrival. We had tried Prairie and Sierra, Custer and Cheyenne (for their South Dakota Heritage), and Scotland and Rob Roy (for the Grandfather Mountain Highland Games), but none of these seemed appropriate.

The names finally emerged, however. The black female we would name Yonahlossee, after the road that winds up Grandfather Mountain, the present day U.S. 221 from Linville to Blowing Rock. Yonahlossee is a Cherokee word meaning, "the trail of the black bear". The cinnamon male we named Kodiak after the brown bears on Kodiak Island,

Pam Bottle Feeding Kodiak

Alaska. It was certainly a big name for a tiny cinnamon colored black bear to live up to! We were very excited about raising Kodiak because Grandfather Mountain had never obtained a cinnamon black bear. Our Kodiak, however, is *not* a Kodiak bear. Kodiaks are the largest of the grizzlies or brown bears, a different bear species. Our bear was named after them solely because of his color, because less than one percent of black bears in the eastern United States are brown. What a great way to educate the public, because many people do not realize that black bears can be blonde, brown, bluish, and even white!

It took the cubs a while to adjust. We kept them in the office for the first month, with trips into the cub habitat several times a day. The public enjoyed the bottle feeding sessions (the formula consisted of goats' milk and baby rice cereal), but the keepers all had battle wounds.

Bear cubs are cute, but they're definitely *not* cuddly. They have terrible tempers, and if the milk runs out of the bottle and one doesn't have another ready--keeper beware! They scream, they claw, they climb legs, and they bite. No keeper came out unscathed.

Yonhalossee

Even though we loved them, we were very happy when the time came for the cubs to stay in the cub habitat. Sharing a small office with active, feisty bear cubs that demanded their way all the time was a bit trying. The cubs liked the new arrangement, too. They loved climbing in the trees and playing in the grass and flowers. They were terrified, however, of the adult bears in the next habitat.

The adults couldn't reach the cubs through the chain-link fence, but they could frighten the cubs by charging the fence if the curious cubs got too close. When this happened, the poor defenseless cubs would climb the first tree available and scream at the top of their lungs. One day, it took us an hour and a lot of patient coaxing with peanut butter and watermelons before the cubs ventured down out of the tree after such an incident. But when they saw one of the adult bears poised on the opposite side of the fence, they scurried right back up again. Fences, at least to bear cubs, are not barriers.

As summer evolved, the unsteady swats of playing cubs became boisterous chasing matches, and the cubs wanted the humans to participate. Humans, unbeknownst to bear cubs, have very thin skin, and the playful bites they bestowed on each other, then delivered to us, left everyone with bruises and broken skin.

Kodiak was the worst. When play erupted, and the keeper couldn't get out of the way in time, Kodiak would turn on the keeper, his eyes glowing and his ears pressed to his head. He would jump (his favorite launching place was the hollow log) and land on the keeper with all claws wrapping around a leg and jaws chomping down. Pushing him away only prompted him to jump again, and the only way to get away from him was to run for one's life and hope that Yonahlossee diverted his attention. The visitors loved Kodiak's antics, much to the chagrin of the poor keeper.

Film crews from different television stations came to capture the cubs' charm on video, only to discover that cubs loved knocking over tripods to eat cameras. The cameramen themselves were wonderful targets. The cameraman would reach down to pet one of the cubs, only to find tiny bear teeth clamped on his hand and leg. The man would proudly wear these "trophy wounds" back to the station exclaiming, "I've been bitten by a bear!"

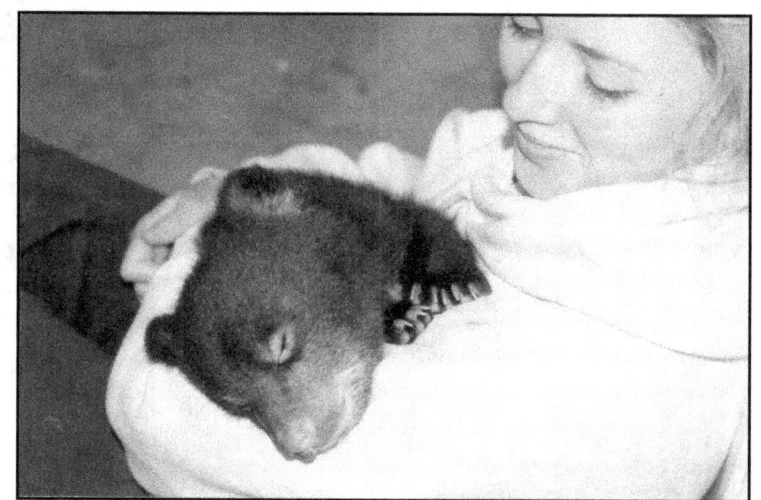

Rene' and Kodiak

In the fall of 2000, these two bears took their place in the big bear habitat along with Carolina, Dakota and Maxi where they continued to provide visitors and their keepers with endless enjoyment and memories.

Critter's fawns   Photos by L. L. Mitchell

CHAPTER 10
THE DAY FROM HADES

Critter, one of the does, is limping at morning check. It doesn't take long for us to figure out why. Flora, one of the younger does, has everyone stirred up. Her ears are pressed to her head in an aggressive manner, and she's chasing everyone around. Critter has been one of the targets, and in her flight, she has injured her foot.

Taking a long look at the incensed Flora, I understand why. Her once swollen belly is now concave. She must have had her fawns in the night and doesn't want anyone getting close to them.

Entering the deer habitat, Sherri and I search for the newborns, but fawns are a lot harder to find than one would think. We call them nature's Easter eggs because it's like an egg hunt every morning when we check on them. They curl up tightly within the roots of a tree or hide in the tall grass and remain completely still.

Finally, we find two healthy doe fawns on opposite sides of the habitat. Mother deer almost never keep their twins together when hiding them.

Now that we've found the fawns, we turn our attention back to Critter. Her left hind foot is so painful that she doesn't want to put it on the ground. Flora continues to chase her. We will have to get Critter and her two-week-old fawns in the back deer habitat away from the others.

Getting Critter into the back habitat is easy. She was hand-raised by humans and is pretty trusting of her keepers. She also knows that succulent, tall jewelweed awaits just beyond the open gate. Her two little fawns are not nearly as cooperative.

At two weeks of age, they've already found their spindly legs to be good escape tools. Instead of hiding in the tall grass like they once did, they jump up and run the moment we approach. Chasing them outright does no good. They're entirely too fast for their two legged adversaries. We realize we're going to have to corner them against the fence to catch them.

The other deer, confused at what we are doing, begin running too, getting in our way. Finally, I corner one of the fawns along the fence. He runs down the fence line, trying to reach Critter on the other side. I'm about to reach out to grab him when Flora's newborn decides to get up and staggers into the older fawn's path. Critter's fawn jumps clean over her and continues running for his life.

The newborn fawn sees me approaching and staggers forward a little, trying to get out of the way. Her legs are not working quite right yet, and she pulls the only defense she knows. She goes completely limp and falls to the ground with her long legs splayed out in all directions and remains motionless. She thinks she's hiding.

I finally catch Critter's fawn in the corner of the habitat. He kicks and kicks, and begins the dreaded "distress" cry. All the does run to his aid as he cries and struggles. I get him through the gate to his worried mother before the others become too upset, then look around for the other fawn.

Critter's second fawn is standing up at the pond, his head and ears erect. We begin the chase. It has to be a comical sight for visitors as we chase the nimble fawn around and around the habitat. I corner him in the rocks by the pond and catch him as he leaps into the air to run past me.

His cries of terror fill the air, and once again, the does rush over. I run down the hill toward the gate with the kicking fawn as Star, Heidi and Flora pursue me. For one brief moment, I think Star might attack me. At last, I thrust the fawn in with Critter and everyone quiets down.

It's only nine in the morning, and Sherri and I are already exhausted. We hope the rest of the day will be a little easier. Little do we know that we are just beginning the Day from Hades.

After cleaning dens and getting bear food ready, Sherri and I feed Mumbles, Gerry, Maxi and Dakota in the Big Bear Habitat, but can't find Carolina. I'm not too concerned, however. At morning check, we'd seen her ambling along the rocks by the pond.

We feed Elizabeth and Punkin in an adjoining habitat, thinking Carolina will appear, but she never does. Sherri spots her way down the steep hill in the right corner of the habitat along the fence. She's trying to dig under the fence to get into Elizabeth's and Punkin's habitat!

"Stop that, Carolina!" I yell. She ignores me.

Irritated, I start down the fence line on Elizabeth's and Punkin's side to see what kind of damage Carolina is doing. When I reach Carolina's excavating site, I'm horrified to see that she's almost through the hole.

"Stop it, Carolina!" I shout again, placing a foot on her head in desperation and trying to push her back to the other side. Her head and shoulders are just about through. If she claws her way into Elizabeth's and Punkin's habitat, they are likely to hurt her.

Desperately, I look around for a large rock or anything that I can stick in the hole to keep her from coming through, but there's nothing within my reach.

Carolina Acting Innocent

Sherri appears through the foliage, clinging to the fence and working her way down the steep embankment toward me. Her mouth drops open at Carolina's accomplishment.

"Sherri, I need some rocks," I say urgently. She looks around, but the only rocks around her are huge boulders that are embedded in the ground. Just then, Mike blurts over the radio to let me know he and Dana have arrived.

I'm so flustered, I don't even tell him what's going on, just that they are needed, pronto, in the lower end of the Big Bear Habitat. I see some promising looking rocks in the yearling habitat below us and tell the two to come through that way.

Finally, I see Mike and Dana running up the hill with Kodiak and Yonahlossee following. "Get those rocks!" I yell, pointing to the large rocks around them. "Throw them through the gate in here!"

By this time, I've managed to draw a rotten log over and place it in the hole, but that won't deter Carolina for long. Now Dakota has joined Carolina in her attempts.

Mike and Dana begin throwing rocks through the gate while trying to keep Yonahlossee and Kodiak from sneaking by them. Sherri climbs down the steep hill to help, hanging onto the fence to keep from tumbling through the clinging blackberry vines.

Good old Gerry the bear comes to our aid at that moment. She runs down the hill and chases Carolina and Dakota away. Normally we disapprove of Gerry's habit of chasing the two younger bears, but this day we are grateful.

We work as fast as we can, but carrying large heavy rocks up steep inclines through scratchy blackberry brambles while clinging to the fence is extremely hard to do. Finally, we feel as if we have enough rocks on that side of the fence. We climb up the hill to the entrance of the big bear habitat and start back down on the other side, intending to load rocks into the hole from that side. Carolina and Dakota are already back at the hole trying to get the first load of rocks out!

Dana distracts them with peanuts and lures them to the front to keep them entertained while we find large rocks to put in the hole. After we're satisfied with our handiwork, we feel faint from fatigue and hunger, but we are far from through. After lunch, we will have to concrete those rocks in place to keep those kooky bears from digging through.

I suddenly wish Tanya were here instead of me. She is the "McGyver" type and likes to be. (McGyver was a television show in the late eighties about a man who could fix anything with the least amount of essentials). I have no desire to be McGyver and absolutely hate pouring concrete. But sadly, I am here, and she is not.

After mixing concrete, water, sand, and gravel together in a wheelbarrow, we have enough mixture to finish the job. We take the truck around through the service road with the weighted down wheelbarrow full of sloshing liquid cement and drive up to the big bear gate. Mike pours the cement in buckets and we struggle to get them into the habitat.

Even though the buckets are only a third full, they feel like they weigh a ton. To top it all off, Carolina and Dakota, seeing the buckets, think they are food buckets and come to investigate. Dana once again persuades them up front with treats as we make our way down the hill with a bucket in one hand and a handful of fence in the other. Hauling rocks is nothing compared to hauling concrete. It is a most arduous experience.

Finally, the concrete has been poured and we sit back, red-faced and breathing heavily, sweat pouring from our brows. I feel as if I'm about to collapse, but it's a good feeling to know that Carolina and Dakota cannot get into the other habitat. Punkin and Elizabeth are very aggressive to other bears, and if the escapees had succeeded, they might have been severely injured.

When I get back to the habitat office, I look at the clock and smile. It is three-thirty ... almost time to go home. I can already imagine myself in a nice hot bubble bath soothing my aching muscles and scratched up legs. Before I can go home, though, I have to give Critter a penicillin shot. She has acquired an infection after the birth of her fawns.

As Sherri and I walk toward the deer habitat with the new dart gun that I don't really like, I notice Mike on top of the otter display. He's standing there with his mouth wide open and is obviously staring at the swimming otters below him.

"You won't believe what just happened!" he said excitedly. "I was trying to catch a garter snake and it jumped into the otter pond!"

"What?" I ask, scrambling up to stand next to Mike.

"When the snake hit the water, the otters were all on it like piranahs!" Mike said. "They took it to the bottom in a death roll!"

I glance into the underground otter viewing area at the visitors' horror-stricken faces. The only thing left of the snake is the tail. Manteo is busily getting rid of the evidence.

Fish, fresh water mussels and clams, insects, crayfish, baby beavers, chipmunks ... and now snakes, I think to myself. Another otter delicacy to add to the list.

" Hey! What fell in the water? Let's go check it out!"

I shake my head and approach the deer habitat. I want to give Critter her shot so I can go home. I load the dart into the gun barrel and attach the pressure gauge. Then I aim and pull the trigger. Nothing happens. Critter's ears prick forward, and she stares at me wide-eyed, realizing what I'm trying to do.

I can't figure out what's wrong with the gun and try to adjust the gauge. When I do, the gun goes off and the dart shoots through the chain link fence into the dirt. Luckily, the medicine doesn't discharge.

I look at Sherri in exasperation. I have to go out of the habitat, onto the service road and through those lovely blackberry vines to retrieve the dart. I walk wearily all the way back around and load the gun again.

Critter has figured out what's going on by this time and tries to get away from me, but I corner her and fire again. Once more, the gun doesn't go off when I pull the trigger. I jerk the barrel off and look for the pink tail of the dart, but just enough air has escaped from the carbon dioxide cartridge to push it to the middle of the two-foot long barrel, and I can't see it.

The dart is supposed to be positioned next to the pressure gauge, so I need to get it out to re-position it. I put the barrel down and look for something to retrieve it with, and when I return with a skinny stick, which turns out to be too big, I realize that I don't know which way the dart was inserted. Both barrel ends look exactly the same. Even if I wanted to shoot it now, I wouldn't know if the needle end or the tail end would come out first.

I'm so exhausted, I can hardly think. Sherri goes back to get Dana to replace her and asks Dana to get a wire so I can get the dart out, along with a syringe of medicine. If the gun won't work, maybe we can just catch Critter and inject her with a needle and syringe.

While I wait for Dana to arrive, I lay flat on my back in the tall, waving grass. I cannot believe how tired I am.

After what seems like an eternity, Dana appears with some wire, and I retrieve the dart. I test the trigger part of the gun before putting the barrel on, and it seems to be working. I put the dart in the barrel, aim, and fire. Again, nothing happens. I try to adjust the pressure gauge on the gun, and the dart shoots out through the chain link fence again. I feel like breaking the gun. I don't remember ever being this angry or this tired!

I decide to use the syringe. I catch Critter, and Dana tries to inject the needle but every time she tries, the needle bends. Finally realizing the needle is too small, I release my hold and Critter bounds away. Wearily, I walk out of the deer habitat and along the fence to retrieve the dart that still has not gone off.

I suddenly get the brilliant idea that maybe I can use the dart needle on the syringe with the bent needle. After that thought, my brain shuts down. It's so tired that it won't work anymore.

Forgetting the contents of the dart are under pressure, I pull off the dart needle and penicillin shoots all over me.

I realize at that moment, I have to go home. Shot or no, Critter will survive. After trying to give the deer a shot for an hour and a half to no avail, I drag myself into my Jeep and head home.

The next day, Tanya informs me she's noticed the trigger of that dart gun has been sticking sometimes and needs to be pushed back into position before the pressure gauge can be adjusted.

And I thank her for that most helpful tip.

Flora, Fauna & Merriweather   Photo by L. L. Mitchell

Sheaba waiting for her fishcicle treat   photos by L. L. Mitchell

CHAPTER 11

FISHCICLES, FRUITCICLES AND OTHER FORMS OF ENRICHMENT

My eyes sweep over every nook and cranny of the cougar habitat. Squeak, Sheaba, and Mina are hiding. Cougars are very elusive, even in captivity, and ours are no exception. Out of all the animals in the habitats, these cats are the ones visitors are least likely to see.

I decide to whip out one of their favorite treats to get them out of hiding ... a fishcicle.

When I come back to the overlook with the bucket, Mina is staring at me from behind several rocks in her habitat. I can just see her eyes and ears. She blends in so well that several of the visitors can't see her even though I'm pointing at her. I bump the bucket against the wall and call her.

Mina bursts through the undergrowth and runs frantically up front trying to reach her treats before the others. Her penetrating yellow eyes stare at the bucket. When I throw the fishcicle to her, she immediately pounces on it, then settles in the grass, licking and trying to pull the fish out of its icy prison.

It will keep her busy for hours and keep her up front for the visitors to see. Sheaba appears next, then Squeak. Mina, the queen of the cougar habitat, hoards all the fishcicles until she's full, then Sheaba and Squeak get their turns.

Not only do the cougars enjoy fishcicles, they also like to play with their bungee rope. Beth has made a rope that's tied to a tree out of reach of the visitors, and during the day, one of us dangles the rope down and wiggles it. Just like a house cat after a string, Squeak will bound up, playing and jumping after the elusive rope.

We also present problem-solving activities to the cougars by putting fish in hollow logs or in a tire hanging from a tree. We sprinkle scents like fox urine around the habitat for them to investigate.

The otters like scents, too. But their favorite things are cat food treats and icicles thrown into the water. They also love the little green ball we throw into their midst. One of them will balance the ball ahead of her on her nose as she pushes it through the water with the others chasing after her. Then she will grab onto it tightly and do somersaults over and over.

Otters never seem to get bored. If we don't give them anything to play with, they find things to amuse themselves--playing in the waterfall, digging in their enclosure, making slides in the snow in the winter, and just playing with each other. When we're working above their habitat raking grass or watering the rhododendrons, all we have to do is swing the rake or hose around and they're off, playfully running and diving in and out of the water.

Peak a boo!

The bears like fishcicles, too, and fruitcicles as well. We mix fruit juice and water, then put peaches, strawberries, blueberries, and blackberries in the mixture and freeze it. Scattering their food and putting it in holes and under rocks also stimulates them to forage like they would in the wild. Because the habitats are natural, the bears also enjoy the blackberries and jewelweed that grow there.

Jane & cubs enjoying a fishcicle

Deer can get bored, too, so we scatter peanuts, apple pieces, raisins and lettuce around their habitat for forage. In rutting season, The Count enjoys battling PVC pipes hanging from a tree; in fact, he battles with every fallen branch, the fence, and even the Woodchuck Habitat sign during this time.

In the fall, we give the animals pumpkins. The cougars love to bat them around, especially if they're smeared with sausage or hollowed out with feline diet inside. Ditto for the bears. The otters treat their pumpkins like they do their little balls, rolling them around and around in the water.

Enriching captive animals' lives has become very important in recent years, and the staff at Grandfather Mountain tries to keep up with the latest techniques. These activities stimulate the animals and keep them from getting bored. They also provide the visitors a lot of enjoyment as they watch the animals jump for bungee ropes, swim after little green balls, or enjoy refreshing fruitcicles.

Carolina & the Great Pumpkin

CHAPTER 12

MILTON THE BEAR

A group has called from the museum and wants a tour, and Tanya volunteers to give it. The group is made up of eighth graders--a little too old for Milton the Bear, I think. But what the heck!

Beth (my escort) and I start out after I dress in the bear suit. Beth is there to keep me from running into things because I can't see too well in the costume.

Tanya has finally gotten the thirty or so eighth graders to quiet down and is talking to them about the deer when I burst into view. The kids don't see Milton at first as I dance in their direction.

Some of them turn and look at Milton. Immediately, I began doing the "moonwalk" dance for them. Their eyes widen and they start laughing. Before I know what's happening, they're running in my direction.

I hold up my hand for them to "give me five" and start dancing again as I start up the path. Tanya frowns at me. "Thanks a lot for interrupting the tour," she says. "I've totally lost control."

I shrug my shoulders and continue dancing toward the cougar habitat. Several of the children follow me as if I'm the Pied Piper. Tanya manages to collect all the kids eventually. I was wrong about eighth graders. They love Milton!

I like to visit the cougar habitat when I am Milton because Squeak hates Milton and his reaction is hilarious. Today is no exception. As I peer with my big Milton head over the side of the wall, Squeak sees me from the back of the habitat. His eyes fix intently on Milton's face. Squeak stalks out of the jewelweed, his ears pressed to his head. He hisses menacingly. I'm not supposed to make any noises inside the costume, but I can't help myself and start laughing.

Squeak Gives Milton the Evil Eye

Pam agrees to be Milton when I return to the office. She's the only habitat employee who hasn't tried on the costume. She dresses in everything but the head, and we start toward the habitats. At the last minute, she adjusts Milton's head into place, but as soon as the head is placed between her shoulders, she throws it off.

Milton's head goes rolling into the woods, and she looks at me wide-eyed. "That head makes me claustrophobic!"

I retrieve the head and put it back in place, and once again, she throws it off. She's about to panic. "I don't think I can do this!" she says.

I pick up Milton's head a second time. "We're too close to the trails for you to not have the head on," I say. "Some little kid is going to see you!"

I put the head back on Pam and lead her into the habitats. She's not allowed to talk (because you are not supposed to talk if you are Milton), and she can't throw the head off because visitors are milling around her. I lead her down the hill toward the deer habitat but halfway there, Pam grabs onto the rail, and I can't make her let go.

A large group of children heading up from the deer habitat spy Milton and runs toward him. They grab Milton's arms, laughing, and pull him around. Milton clings tightly to the rail with both hands and won't allow anyone to pull him off.

I chuckle as I get the children's attention with the candy bag. They leave poor Milton, who is about to faint, and their minds turn to the candy. They meander away.

At this point, Pam doesn't care if Milton is not supposed to talk. "Get me back to the office," she exclaims. She sounds as if she's in a well. "I'm hyperventilating!"

I lead the shaky Milton back up the hill. We're barely on the service road when Pam throws off Milton's head again. Her hair is sticking in all directions. She's red-faced and sweating.

"I'm never going to be Milton again!" she gasps.

Later in the day, I decide to be Milton once more. I enter the habitat running like a rhea, a large flightless bird from South America that looks sort of like an Ostrich. This is the funniest thing Milton does. He is an expert at it. Rheas sometimes get excited and run erratically around with their heads bobbing from side to side. If you've ever seen one of these birds, you can't help but laugh.

I stop that antic and begin to skip down the path with Rene' by my side. Suddenly, my feet get tangled in something, and I fall down. Slowly, I grab the rail and pull myself up, confused, as Rene' starts laughing at me. I look at her through the eyeholes and notice she is pointing at my feet. She's laughing so hard she can hardly talk, and the visitors are laughing, too.

I bend way over to look at my feet, holding onto Milton's head. One of Milton's suspenders has broken and his green shorts have fallen down around his ankles! I begin laughing so hard that I start crying. Rene' tries to pull up Milton's shorts. She tugs and tugs, trying to get them back into position, but they promptly fall back down again. As Rene' escorts me back to the office, I have to hold up Milton's shorts with my right hand while I hold onto Rene' with the other.

And another day with Milton the Bear comes to a close.

RED SQUIRREL   PHOTO BY HUGH MORTON

CHAPTER 13

GROUNDHOGS ARE NOT HEDGEHOGS!
AND OTHER EXPLANATIONS FOR VISITORS

"Look, Daddy!" I heard a little sandy-haired boy say. "What kind of animal is that?"

I was standing behind the father and son at the deer habitat, and I looked up, as his father did, to the object of the discussion. It was a little red squirrel, busily eating an acorn on top of a flat rock above the deer pond. The little boy's father didn't answer, and when I glanced over at him, I could see the man was struggling. And then he said those three dreaded words: "I don't know."

My mouth dropped open, and I glanced up at the squirrel. I could almost swear that it, too, was staring at the father with exasperation. I could possibly understand if the man didn't know what *kind* of squirrel it was. After all, red squirrels are found only at high mountain elevations and this man was possibly a flatlander. But surely he must know it was *some* kind of squirrel. That little animal above us had a bushy tail and was eating of all things, an acorn! I couldn't let that one go, and that father and his little boy now know what a red squirrel is.

It's perplexing to us habitat employees that people don't know their animals. I began learning about animals from the time I could walk, and I sometimes assume people know things about animals that I take for granted. But they don't, and that's where the habitat staff members step in. They try to teach visitors how special nature and its animals are. They also try to correct them politely if they overhear such conversations as the one described above.

Most people live in cities now. They don't see wildlife and forests that their ancestors did. Through all those generations, in this era of computers and modern technologies, nature gets put on the back burner. Nature becomes a place people love to visit, but they don't really know enough about it to appreciate it. Even if they do appreciate it, they sometimes don't know what it is that they are appreciating!

The habitat employees are here to make sure that the visitors to Grandfather Mountain know what they are seeing. They want them to enjoy and learn about the animals in the exhibits and the ones they might see scampering around the trails or backcountry. I hope the following will help. To add a little humor, I'm including visitors' perceptions of what these animals are and some frequently asked questions.

"Hey! Look at that mouse!"

"Well ... you have the right order, Rodentia. This animal is really a chipmunk. Mice and chipmunks are both rodents, but they look totally different. Chipmunks have stripes. Mice don't.

Mouse

And mice are smaller and have long, almost hairless, tails.

Chipmunk

"Daddy! Look at the penguins!"

Thank goodness a child said that. He was talking about the bald eagles!

Bald Eagle   photo by Hugh Morton

Groundhog   photo by Hugh Morton

"Your otters have escaped!" or **"Look at that hedgehog!"**

Two strikes. This animal is neither an otter nor a hedgehog, and is not related to either one. Otters are Mustelids and are related to skunks, ferrets, and weasels.

Hedgehogs are insectivores and are not even native to this country. You are looking at a groundhog or woodchuck. Mountain people call them whistle pigs. Groundhogs are rodents and are related to chipmunks and mice. Oconee (the otter on the right) would not appreciate being called a whistle pig!

Oconee the Otter

"Another name for a cougar is jaguar," said one visitor to his girlfriend. Not true. Cougars have many names including mountain lion, catamount, red tiger, puma, and panther, among others, but jaguar is not one of them. Cougars are tawny colored cats that are found from Canada to South America. They do not have spots and there has never been documented proof of a black cougar. Jaguars are found in Central and South America and Mexico and are rarely seen north of the U.S. border. They are a yellowish-brown color with black spots or rosettes. Jaguars do come in a melanistic or black color phase, but even then, spots can be seen through the jaguar's black coat.

The jaguar is classified with the large cats (*Panthera*) and is the only large cat in the Americas. Its scientific name is *Panthera onca*. The cougar is classified with the small cats (*Felis*) and is the largest of the small cats. *Felis concolor* is the cougar's scientific name.

"You mean ... the antlers on that deer come off every year?"

It's amazing to me how many people don't know this. All animals in the family Cervidae, such as deer, elk and moose have antlers that shed every year. The process starts in the spring. Small buds appear where the old antlers fell off. The buds have a skin covering called velvet. Underneath the velvet, blood vessels and tissue bring nutrients to the forming bone of the antler. The antler continues to develop and branch throughout the summer. In the fall, the blood vessels and tissue around the antler begin to die and the velvet dries up. The buck scrapes his antlers on the trees and branches to rid himself of the

velvet. After the velvet is shed, only the antlers remain. Bucks use the antlers during the winter breeding season to fight for territory and females. As spring approaches and the breeding season ends, the buck's testosterone level drops. When this happens, his antlers are shed and the process starts all over again.

Antlers and horns are different. Animals with horns include the Bovids, such as bighorn sheep, bison, and mountain goats. Horns are made of keratin, not bone. A person's fingernails are made of keratin. Horns are hollow, usually don't branch, and are filled with blood vessels. They are never shed and stay on the animal its whole life. Antlers are made of bone and usually branch.

Photos by L. L. Mitchell

"Did you release all these birds in the habitats?" asked a visitor pointing to a blue jay in the deer habitat.

No, we don't release any birds. They are found naturally on the mountain. Some of the birds stay year round such as dark-eyed juncos, blue jays, goldfinches, pine siskins, red and white-breasted nuthatches, and downy woodpeckers. Many birds are migratory. They're only here during the summer months and retreat to warmer climates in the more southern states or even Mexico and South America. Some of the birds found in the habitats in the summer months are red-breasted grosbeaks, indigo buntings, ruby-throated hummingbirds, catbirds, and rufus-sided towhees. Over one hundred and twenty bird species have been found on Grandfather's slopes during the spring and fall migrations ... and all got here on their own accord!

Morley the Golden Eagle   Photo by L. L. Mitchell

"Are your eagles tied down?" or **"Why don't your eagles fly away?"**

No, our eagles are not tied down. They can hop and walk anywhere in their habitat. They can't fly because they were injured by gunshot in the wild and have had one or most of their wings amputated. Because their wings won't grow back, they were placed in Grandfather's care for the rest of their lives.

Visitors, please keep the questions coming. The habitat staff will be sure to help you if they know the answers. Remember, no question is a dumb question if the answer you receive increases your knowledge! And enjoy your time in Grandfather Mountain's Animal Habitats!

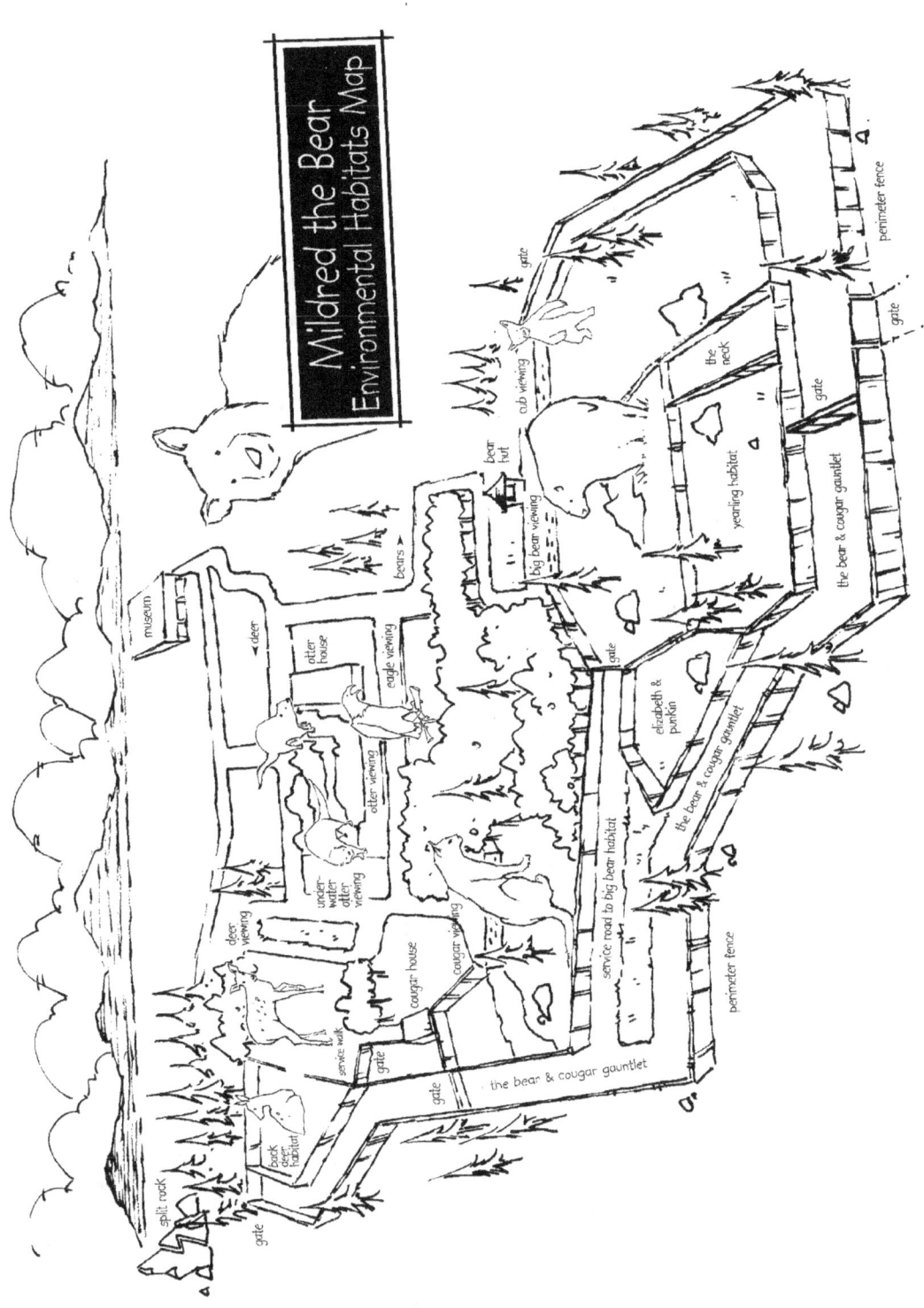

www.ingramcontent.com/pod-product-compliance
Lightning Source LLC
Chambersburg PA
CBHW081328310526
45789CB00018B/2562